The Ultimate Guide to Fire and Explosion Prevention

How To Keep Your Plant From Blowing Up And Burning Down

By Jeffrey C. Nichols

The Ultimate Guide to Fire and Explosion Prevention

Copyright © 2016 by Jeff Nichols

ISBN-13: 978-1530930630
ISBN-10: 1530930634

Visit the Author Website:
www. http://industrialfireprevention.com/

Disclaimer

Acknowledgements

First, I would like to thank my friend and mentor W.B. for convincing me that this book needed to be written, that I was the guy with the experience to do it, and for helping me make it happen.

A special thanks to all of the people involved in process safety who help our workforce come home safe to their families after every shift.

I would like to thank all the great people in my life and career who have helped me along the way, and encouraged me to keep going. To keep helping protect companies and people even when it was hard to convince people to invest in fire prevention, especially when they haven't yet had a fire!

Those people, too numerous to mention each one here, include my close friends and family, current and former bosses and colleagues, customers, industry experts and partners, publishers and media, Authority's Having Jurisdiction, regulators, insurers, and even competitors have motivated me to keep going doing what I love to do. And that is helping companies keep from blowing up and burning down. If you know me, then you know who you are; and I thank you from the bottom of my heart. I could not have succeeded in this business without you!

And thank you the reader, for taking the time to read this book. I appreciate you. I have spent my entire career providing safety systems to the process industries, and trying to educate people to the dangers of combustible dust. So I applaud you for your interest in reading this book.

My customers are my hero's. Those who fight the good fight to keep their plants and people safe, day in and day out. You have my admiration. You are why I do what I do.

My purpose in this business has always been to leave it safer than I found it. Hopefully, I have done my part to make our industry safer.

Make it Safe,

Jeffrey C. Nichols

Contents

Part 2
The Process

Part I

Background

Introduction

Introduction

"How do we keep our plant from blowing up and burning down?"

"Why do we have fires? Aren't they just a normal part of the manufacturing process? What is combustible dust? Why is it so combustible? So explosive? Are all processing dusts combustible?"

"What can I compare it to, or how can I explain it to my colleagues and employees so they "get it" and want to help make our production safer? How do I help them understand the kind of risk we are dealing with every day in our plant? So that they understand how dangerous a combustible dust fire or explosion is?"

"How do we keep our production and people safe, and apply the prescriptive safety standards specifically to our own unique process? How do we become compliant?"

"And how can we do it without having to hire some high priced PhD, consultant or trainer who uses big words, talks over our head and views everything from the 30,000 foot level? How do we make it real and relatable to everyone who works at, or comes into our plant?"

These are the types of questions I hear, and I see in print, about the combustible dust issue. And we are not just talking about safety for your employees, supervisors, and management, but also vendors, visitors and investors. Everyone who works at or visits your plant. The "Stakeholders" as consultants, lawyers, engineers and the media would say.

This book will attempt to answer these and other questions you may have about combustible dust, the causes and mitigation of fires and explosions, as well as compliance. And do it in such a way as to make it easily understood and relatable by even the newest employee. After all, safety is everyone's job.

Preface

This book is the layman's guide to practical prevention of combustible dust fires and explosions in the combustible dust processing industries.

I wrote this book for two reasons. First, there is no other publication on the market written in laymen's terms for employees, the front line worker, for maintenance and electricians; for supervisors and managers, to help understand and explain the dangers of combustible dust and how to mitigate these dangers.

This book is written for the average Joe that does not have access to, or the time to search through a lot of research and data, who could care less about equations, formulas and science language found in most of the books and articles on this subject. This book is written for people who just want the simple truth, and simple, easy to understand solutions. Those who just want to know how to prevent fires in their facility, and keep people safe.

I Have Been Where You're At

I have worked on the production line. I have had to fight fires at a manufacturing plant. I have had to inhale noxious smoke and I voluntarily put my-self at risk to save another man's business. I have witnessed the devastating aftermath of fires and explosions. I have interviewed people who experienced those horrors first hand, and still have nightmares about them. There is no other general publication written on this subject for the front line manager, and other employees that are not fire protection engineers or safety professionals. Although the engineers and safety people reading this will also get a new take on combustible dust fires and explosion mitigation from a different and unique perspective.

Also, this book is written for real people. It is not intended to compete with the scientific fire and explosion protection publications written by PhD's and explosion propagation scientists, that are typically written for other experts. Nor is it meant to compete with the existing codes and standards, only to compliment them in understanding from a lay person's perspective.

I'm Not Perfect, Nor Is This Book

This is not intended to be a perfect book or provide perfect advice, because I'm not perfect, and I do not have perfect knowledge, nor do I claim to. This layman's guide is based on my own decades of experience in helping protect the process industries from fires and explosions.

And I want to share any knowledge I can that will help educate and protect more people. It is as simple as that.

Safety Must Come First

I am writing this book from the heart in a conversational tone, to the guy working in the mill. For the express purpose of sending him the message that safety must come first. That people's lives and lively hoods depend on it. That complacency kills. That combustible dust is as dangerous as gasoline, and should be treated as such. I know who you are, the challenges you face, the dangers you work around every day, many times not realizing how dangerous this dust is that is everywhere in your plant. You are dedicated to your job, your employer and your family. And I am dedicated to helping keep you safe, so you come home to your family after every shift.

Some of the chapters in this book were transcribed from interviews, articles or presentations, and there is some redundancy which I left in for reinforcement of the ideas presented. It is not perfectly written, nor is it meant to be. The whole purpose is to get a point across, not to provide perfect punctuation, or scientific equations explaining how combustible dust is tested in the laboratory. This is not textbook theory. Understanding the principles of combustible dust fire and explosion protection may save your life, and those you work with.

Real World Advice

This is real world advice for real people, based on real world experience and application. Period.

Understand this is the first draft. The next draft will likely have more data, more examples, more proof reading, and better editing. I thought it more important to put out good actionable information, and to get it out as quickly as possible. But I also believe in continuous improvement, so future editions will provide a better product for those concerned with perfection.

Chapter 1

What is Industrial Fire Prevention?

A Compelling Reason for Safety!

People and companies generally do not invest in fire prevention without some compelling reason. Most often a fire, or worse. Sometimes a governing body or authority having jurisdiction, or insurance company will drive safety.

Our Mission, Our Purpose, Our Cause

At Industrial Fire Prevention, we are in the fire prevention problem solving business. We help save manufacturers from burning down and blowing up! And we help save lives. That is the only thing we do. It is our mission, our purpose, and our cause.

We specialize in providing various types of safety and hazard monitoring systems, fire prevention systems, fire protection, and explosion protection systems to the combustible dust processing industries.

The basic premise being that if we can identify the risk early, we can help prevent combustion and help prevent fires. And if we can help prevent fires, then we can help prevent explosions in these processes. Thus, we help protect people, production, and manufacturing plants from the devastating effects of fires and explosions.

Fire prevention in the combustible dust processing industries consists of a series of strategic steps in evaluating risk, recognizing hazards, defining objectives and a strategy for managing these hazards on an ongoing basis, as well as making decisions and allocating resources, and implementation of the various engineering and administrative controls to produce layers of protection.

Each layer of protection then acts as a firewall to prevent the spread of a fire or explosion to other parts of the process or to other processes.

This layered protection system design is based on several levels or layers of hazard monitoring, fire prevention, and fire and explosion protection systems.

These systems may be different for different industries and different applications. Each application is unique.

For example, in the grain industry the most common systems used to detect ignition in the incipient stage are bearing temperature, run time, and speed monitoring, along with belt alignment and plug up detection, as well as silo temperature monitoring. Many times these hazard monitoring and safety systems can also provide trend analysis for preventative maintenance.

In the building products and wood products industries, as well as the paper and pulp, the textile and non-woven industries the most common prevention system is the infrared spark detection and extinguishing system, which detects glowing embers and hot particles. Spark detection systems are designed to detect and suppress sparks in the incipient stage, thus preventing ignition and fire.

Other industries might use combustion gas monitoring, aspiration smoke monitoring, temperature monitoring, flame detection, or other types of equipment and systems we will discuss in more detail in the coming chapters.

The benefit of working on projects in many various industries over several decades, is cross industry information sharing and safety system application, where we can apply safety systems common in one industry to help prevent fires and explosions in another industry. Especially when strategically combined with other types of hazard monitoring and safety systems.

We can virtually monitor most if not all hazards in the process. For example, spark detection at the processing equipment, mechanical and pneumatic conveying systems as well as the dust collection systems, bearing temperature, run time and speed on conveyors, all the way out to emissions monitoring at the dust collector, and everything in between.

Chapter 2

Helping Protect People And Property

Is Safety an Investment or an Expense?

Prevention is the best insurance. In providing our various fire prevention systems to manufacturers, we help protect their manufacturing personnel as well as help protect their production, help them stay in business, and not be shut down for any unplanned and extended periods of time. Downtime kills businesses.

Safety is not an expense, but rather an investment that pays dividends in your process safety, preventative maintenance, optimized production, and in preventing serious down time. A safe and well maintained process runs longer and more efficiently. Your people will be happier; your insurance rates don't ramp up suddenly. Your reputation as a safe employer is safeguarded, and the media and regulatory agencies aren't coming around looking for problems in your plant.

Safety also helps protect your reputation in your industry. If you gain a reputation for having fires and being an unsafe place to work, then you're employees don't feel safe, your managers don't sleep well at night knowing their people aren't protected, that they may get a call in the middle of the night that the plant is on fire. And you will be on the radar of regulatory agencies.

OSHA starts breathing down your neck. Insurance rates start going up. Insurance companies actually cancel companies, or even entire industries!

Insurers actually pulled out of the wood pellet industry in Canada for a period, until they got it under control, because they were having so many fires and explosions. At the same time workers compensation insurance went up 50% in 3 years in that industry. As industry professionals, we must self-regulate. As manufacturers, you must protect your plants and people. You are responsible for a safe work environment. If you fail to protect your people, then government will surely step in to help you, at a steep price.

Navigating the Road to Safety and Compliance

Navigating the OSHA, NFPA, FM Global codes, standards and best practices can be confusing and complex. But don't worry we are here to help you.

Every day for more than the last 20 years, I have worked with small businesses, as well as large multi-national companies and everyone in between, across the country who are engaged in the same battle as you.

Yes, I call it a battle. You are battling competitors, vendors, customers, employees, insurers, government regulators, the economy, and sometimes your own family members, just to stay in production, to stay in business.

I'm on your side. My team is on your side. My partners and network of associates is on your side. My goal is to help you make your manufacturing process safe. One thing I must tell you is that your current level of safety is the result of your decisions and actions over the last few years. And maybe, just maybe you have also been lucky.

We are here to help you. To give you the shortcut to compliance and safety. The shortcut is always the shortest and fastest route. Sometimes the shortcut may look confusing. But we are here to help you navigate the road to safety and compliance in the shortest time possible.

Chapter 3

My Role In Fire Safety

Businesses Protected, Injuries Prevented, and Lives Saved

I do not know how many lives I have saved. Or how many serious burn injuries I have prevented, or how many fires were prevented because of my work. Such is the life of a safety professional. I do not know how many companies I have helped stay in business, protected their people, production, and reputations in industry; that would have been shut down or destroyed had it not been for my help in protecting their facilities.

However, I do know of several companies that did not install the systems I recommended, that ignored my proposals and ended up having explosions that killed people, had lawsuits and ended up going out of business. These cases are tragic. They could have been prevented. These lives, jobs and companies could have been saved.

If I were to read the alarm history of all the hazard monitoring, spark detection, fire protection, and explosion protection system installations I have sold over the years it would be hundreds of thousands of events we have successfully prevented. And that is what drives me.

Most people can measure progress in their job or business. Selling fire and explosion prevention systems and not knowing exactly to the extent I have been successful is somewhat confounding. But I do have the deep satisfaction of knowing all of the plants and people I have played some small part in protecting. There are hundreds of customers I have helped protect over the years. And that gives me deep satisfaction.

Humble Beginnings. You Have to Start Somewhere!

I wasn't always this smart! I started out in 1979 sweeping floors and doing shipping and receiving at a sheet metal outfit on the west coast, for a new "spark detection" division they had just started up. They fabricated a lot of their own dust collection systems and components, as well as a lot of other specialized sheet metal work.

They had just started importing into the United States some of the first spark detection and extinguishing systems to help protect these dust collection systems from fires, and I was the second employee of that division.

I was responsible for helping set up and test the systems. And I would watch the technician do repairs. Pretty soon I was also going school to learn to be an electronics technician, and started doing troubleshooting of the boards, circuitry and components in my spare time.

There was a lot of opportunity to learn. I was learning electronics, I was learning about dust collection and pneumatic conveying. And soon I was helping start-up and commission the spark detection systems in the field.

We started working with the large building products manufacturers in the Pacific Northwest, designing and building some of the first abort gates and back-blast dampers for dust collection systems. I was responsible for testing the speed of the abort gates and making sure they were working correctly.

We built dust collection and conveying systems for plants all over the country in many various industries. And put spark detection on all of these conveyors and dust collectors, as the dust these manufacturers were creating as a byproduct of manufacturing was very combustible.

We tested and compared equipment from all the major manufacturers of spark detection systems. We also did customer and insurance company side by side tests of these systems. This was a tremendous education in spark detection.

Soon I was trained as a field technician. A young single guy on a company expense account traveling all over the country to new manufacturing plants every week. Seeing how so many things were manufactured, and learning something new every day. What a great opportunity.

It was a great job!

That time in my life was a great learning experience and gave me a real advantage to understand pneumatic conveying, dust collection, electronics, process controls, spark detection, as well as how the various types of spark detection and extinguishing systems performed and protected processes, conveyors and dust collection systems in many different industries.

After a handful of years they asked me to open a sales office in Atlanta. Apparently as a field technician, I was selling more spare parts and service contracts than the other salesmen. I would always spend a lot of time training the operators as well as the maintenance people and electricians who would be responsible for operating and maintaining the spark detection systems. And I would always recommend spare parts to keep on hand to keep these systems up and running should a part go bad. Many of our big customers would not run their production if the spark detection system was down! So it was important they should have the right spare parts and know how to maintain and service the equipment.

Burn Out

In the mid 1980's I moved to Atlanta to begin a new adventure in sales. After several years my sales grew, but my income did not grow in proportion to my sales. I realized, working in a family run company, that I would not really get any significant advancement, nor would I make a lot of money, so I started looking around for other opportunities to go into business for myself, in other unrelated fields. Plus, I was getting burned out. I felt like I was just selling equipment to companies, and having traveled steadily for several years playing the sales game, and several years in technical service prior to that, I was getting somewhat complacent and burned out.

The company found out about it and fired me at the company Christmas party in Oregon!

After that I went to work for a buddy building custom homes, and doing real estate sales, and then into remodeling, and then building homes for a production builder in Atlanta. This also gave me a good background in the building products industry, seeing how these products were used and performed, in addition to knowing how they were made.

After several years, I was contacted by the original spark detection company to come to work in sales for them, as they had broken off relations with the sheet metal outfit, and had opened their own corporate offices in the United States.

I always enjoyed sales as well as helping protect companies, and thought it would be a great opportunity eventually go into business for myself, so I took the job and went to work for them for a couple years, before going out on my own. And so I started over building a sales territory from Atlanta covering the Southeast.

This Time Was Different

But this time it was different, I was different. Over time I began to see the effect my work was having on these companies over the long term. I had engineers, plant managers, and corporate safety managers from major companies recognize me and say things like *"Hey, do you remember the favor you did for me that time? You really helped save our ass"*. Or *"Hey, we haven't had any fires since we put in your system"*. Or, *"Yes, we had a fire in that piece of equipment, but your system did its job and kept us from burning down!"*

These were really positive effects my work was having on these people and their companies. I was doing something of grave importance, which gives me deep satisfaction. That is what keeps me going.

Complacency Kills

It is easy in manufacturing to get complacent with the way things are. The the way things are running now or have been running. I cannot tell you how many times someone has said to me *"We have been running this way for years and never had a fire"*.

But also I cannot tell you how many times I've seen or read news reports about a facility that has been operating for years, and then one day they have a large catastrophic event. People get hurt. Sometimes people get killed. And sometimes people lose their jobs when the plant closes and the company goes out of business.

In the production process, the opposite of complacency is vigilance, and continuous improvement. We must always be improving our process efficiency, and we must always be improving safety, with an eye on protecting our most important asset, our people.

Part of safety is training and awareness. In reference to the combustible dust processing industries, this means awareness and training on the hazards of combustible dust, what it is, and what it can do. How combustible dust fuels fires. What causes ignition and fires, what hazards to look for in your process, and how to recognize the leading indicators that often lead to having fires.

The Fire Triangle

There are three basic ingredients required for a fire to happen in your process.

1. Oxygen,

2. Fuel, and

3. Ignition (Heat).

Most of the time, in the process we deal in, oxygen is going to be present in your process, unless you are inerting the entire process. So, you will have oxygen present in your process in most cases, that is a given.

You also have Fuel. You are making something in your process that the main ingredient, or some part of the mix of ingredients, parts, or pieces are flammable. And the finer and drier the product, the more flammable it is. So as the product moves down the production line, it is being heated, treated, refined, shaped, molded, milled, blended, or processed in some way.

And as the dry, raw product is processed, it creates what are called fines. Fines are fine dust, combustible dust. Every time you move or manipulate the product you have the potential to create fines, combustible dust, which is your fuel, as well as friction and heat.

Combustible Dust is Fuel

Many processes have dust collection or aspiration systems to carry off fumes or dust. These processes will either use a local small dust collector at the machine, or a dust collection system with multiple pickups, or hoods, and ducts going to a remote dust collector and fan, preferably located outside. If your processing plant is of any size you will have a dust collector or multiple dust collectors outside the building to collect this dust. Again, this dust is fuel.

So you have Oxygen, and you have Fuel naturally occurring in your process all of the time. You also have heat present. As your process moves and manipulates the product it is also causing friction and heat. Some processes may also be heating, drying or milling a product. These are all points of possible Ignition.

So there you have it. You generally have all three of the ingredients necessary for a fire in your process. People tell me *"we haven't had any fires"*. This just means you haven't gotten the recipe right - yet. If you have all three of these ingredients in your process, sooner or later you are more than likely to have a fire.

Warning Signs

Many times when I interview personnel after a fire or explosion, and I ask them if they noticed anything unusual leading up to the event, I am told things like *"well, we kept smelling smoke, but couldn't figure out where it was coming from"* (and they kept on running). Or, *"I heard several burps"* or even *"I heard a thump"*. What they are describing are precursors, leading indicators of a fire or explosion, signals that something was happening or about to happen. Often the signals are missed, or ignored, because production must be maintained. If we are to prevent fires, we must begin to recognize the warning signs.

When people tell me they have never had a fire, and I am walking through their plant, often I'm looking for any signs of burns on the equipment, conveyors, ductwork and collectors. I'm also looking for build-up of combustible dust, and dust emissions; as well as potential ignition sources. This tells me if they have had a fire or are about to.

Often I will see a buildup of product or material, combustible dust, on motors, pulleys, belts, bearings, fans, etc. This is combustible dust, *fuel*, layered on a potential ignition source. This is a fire waiting to happen.

But since it hasn't happened before, no-one is expecting a fire to happen now. They aren't yet aware that combustible dust can be as combustible as gasoline or natural gas. Do you think they would let layers of gasoline lay around on the floor, or running equipment? Or leave a natural gas leak unattended while operating production machinery?

Another thing I ask them is "when was the last time you changed the bags, or cartridges out in that collector?" Then I ask them if they have ever seen any pinholes from sparks in their filter media. You would be surprised at how many times people say *"Well now that you mention it, yes we have seen burn holes in our filters"*. This is another precursor to a fire or explosion. Again, they have just not yet gotten the recipe right!

Where We Come In

By doing a very narrowly focused dust and ignition hazard analysis of the process, we're looking for any combustible dust hazards, dust layers, or emissions that might be coming from the process, as well as any possible ignition sources existing in the process, with an eye on how to prevent fires.

Then we're going to provide equipment, safety systems, or turnkey solutions to help protect these various parts of the plant, the various processes, production, and personnel from fires and explosions. We are primarily manufacturers' representatives. But we also get involved in protecting plants on many levels, from just selling equipment, to full turn-key packages, to helping with combustible dust testing, training, and consulting. Along with our technical partners, we have a whole menu of hazard monitoring, fire protection, and explosion protection systems and expertise to choose from based on the problems needing solved and the application, as well as the budget we're working with.

We'll take the best solution, the best equipment, and the best fire protection system for your application based on your particular budget, and then we'll have one of our partner companies do the installation of the equipment or system. They'll also do the startup, the service, and any ongoing maintenance on these systems, if desired. That's basically it in a nutshell. How we do it is, again, hazard analysis and then applying the proper engineering controls and systems to protect the process from fire and explosions.

So, if you think you may have a combustible dust issue, or are concerned about possible fires or explosions, we are here to help you. Our contact information is listed in the back of the book.

We also have an offer in the back of the book, to send this book free of charge to someone on your behalf. If you know anyone who could use this book to help make their process safer, I would like to give this book to them as a free gift on your behalf. See the order form in the back of the book.

Chapter 4

Engineered Systems

Layered Safety Systems

Fire Prevention

Some of the primary engineered systems that we use to help prevent fires, called spark detection systems, are basically looking to detect a fire in the insipient stage as sparks or embers are being created. Because, if we can detect the sparks and embers, then suppress it or divert it out of the system, then we can prevent the fires. If we can prevent the fires then we can prevent explosions. And if we can prevent the fires and explosions, we are also preventing the secondary catastrophic explosions

We can add various types of equipment looking for anything that can create ignition in the process, from bearing temperature, and any kind of heat or spark producing equipment we're going to monitor with spark detection, ember detection, temperature, flame detection, CO or smoke detection, and then we use a variety of different types of either suppression or diverters or both, or some other way to manage that hazard and get it out of the process, primarily without effecting production.

The whole purpose of fire prevention is to detect a hazard, handle it without effecting production, so production can keep running. It basically operates in the background. Then we will add additional layered safety systems, installing additional fire and explosion protection systems to further mitigate any larger events.

Fire Protection

Basic fire protection systems would be your typical sprinkler or deluge systems, or they could be a fire suppression chemical, inerting gas, powder, or fire protection foam, or a combination of these, depending on the class or classes of fires involved. It depends on the process, what you are manufacturing, what the manufacturing process can tolerate, and how can we do it and keep you safe and not effect production unless there's an upset condition. Then, of course, when a serious problem is detected, you've got to shut down production, get people out, and flood and area with some kind of fire suppression.

Explosion Protection

Explosion protection will utilize various types of explosion vents or panels, or possibly explosion suppression, and chemical and/or mechanical isolation systems on dust collectors and other vessels containing combustible dust. There are many various types of each of these and configurations, that we can apply, again based on the application.

Each Application is Unique

Some engineers or product managers at large companies want to standardize around certain types of protection design, and this can be done to a certain extent, but it truly does depend on your individual process, as no process is exactly identical. Plus, you also have to take into account mission continuity of critical processes vs, non-critical processes, as well as life safety in each process.

For example, a large building products manufacture wanted me to give them "my best design" to protect their dust collectors company wide. My point was that no two processes are identical. That I would not protect an MDF mill the same was as I would protect a Plywood mill, or an OSB mill, or a Lumber mill. Each one has its own unique and specialized requirements. Nor would I protect the process and equipment the same. I would not protect a cyclone the same way I would protect a baghouse.

As you might imagine, I would not protect a large abrasive sander the same way I would protect a hammermill. I would not protect a dust collector for fine sander dust the same as one for wood chips. Each process is different, the equipment and risks are different, and the hazards are different. It also depends on the process layout. If the dust collector is outside and the explosion vents face the building or parking lot, then I am also going to take these factors into consideration for protection system design.

They wanted a standardized protection scheme, which I can understand and appreciate the reasons for, but in my experience it always depends on the process and budget involved.

Each plant will also have various design, production, equipment, facility or other obstacles to work around. Each plant and process and their challenges are unique.

While you can make some design choices and standardize safety systems around certain types of processes, over the course of several decades I have found that budget is always an object, and each process is different with different risks and limitations or challenges, and required different equipment, even on seemingly identical plants. Different plant managers, production and maintenance people also run and maintain processes and equipment differently. Which makes every application unique.

Chapter 5

What Industries Are Affected?

The Combustible Dust Process Industries

It's not just any industry we serve. Primarily, it's an industry that will create a combustible dust, or uses a combustible dust in the manufacturing process. For example, it could be a variety of products ranging from agricultural products like soy beans, grain, corn, cotton, coffee, tobacco, and animal feed. It could be other food products like sugar, starch, flower, cereal, cocoa, powdered milk. It could be paper and tissue, or personal care products, and what are called "non-woven" products that are made from paper or fiber like diapers or pads, or even cellulose products and textiles.

It might be the metals industry, or processes used in the lead or metals industries like welding, media blasting, thermal coatings, extrusion, casting, recycling, etc. And a large amount of our business comes from the building products, wood products, and paper products industries, as well as biomass and waste to energy.

For example, in the primary wood industry, we have a lot of customers in the building products business. They're making lumber. They're making beams and trusses. They're making siding, engineered lumber and structural panels like plywood, oriented strand board, fiberboard, and particleboard. Those are a lot of our biggest customers. Even in the secondary wood products industries, we have many customers that make furniture or furniture components, cabinets, doors, windows, molding, or even extruded plastic wood to make decking for homes and commercial use. We protect a lot of specialty products processes like wood chips for animal bedding, there's also charcoal, picture frames, and other arts and crafts products that are also made from wood.

We do a large amount of business in biomass, and what they call the wood pellets industry as well as recycling and waste-to-energy. Basically, as Europe has moved away from fossil fuels, they're utilizing biomass and wood pellets in place of coal in their power plants to generate electricity, and they need a large volume of this type of biomass material as a coal substitute for the large power plants.

As they import a lot of these fuels from the abroad and the US, our company has become known as experts in this field, helping protect these alternative fuel plants from fires and explosions. It's a fairly new industry and one of the first things we had to convince people in this industry was that *"hey, you guys are making fuel!" And the byproduct of your fuel is even finer dust that's even more explosive!"* It's been a learning curve for the U.S. biomass industry.

We work in a lot of other industries including the metals industry where they may be cutting, grinding, polishing, welding, or extruding metal. It could be a foundry; it could be a recycling facility where they're grinding stuff up. It could be media blasting, like steal bead blasting, similar to sand blasting, or walnut shell blasting they use to clean these metals with. In aerospace and other manufacturing, it could be metal coating or what they call plasma spray where they use the high temperature spray and make powdered metals molten and spray on metallic coatings for various aircraft parts.

We also do a lot of work in the rubber, tires and recycling, as well as chemicals and bulk powder, pharmaceutical, and nutraceutical industries even plastics, nylon and vinyl, because even in these industries the finer and drier the powder, the more combustible it becomes.

In the recycling and waste to energy industries we get into some really interesting projects. We've even worked with carpet and nylon where they're grinding up waste and using it for boiler fuel. Many people realize that nylon might catch on fire, but they don't realize that if it's ground up fine enough, it's also explosive. Those are some of the more common industries and some of the major industries we work in.

Every Day is a Learning Opportunity

We also do a lot of other uncommon things that we get into as side businesses once customers are familiar with our work, and we've done a good job for them on other projects. A lot of times they will ask, "hey, can you help us protect this type of process?" Some we have had to invent along the way, such as a recent flame detection and quick burst CO_2 system used for wire extrusion die machines.

There is always something new to learn. Every day new challenges.

If you have challenges with sparks, embers, fire and explosions, feel free to contact us at the addresses at the back of the book. We are here to help.

Chapter 6

Why Do We Have Fires?

Refining Products Causes Fires

As we mentioned previously, along with the fire triangle, the reason these manufacturing companies are having fires is because they bring in a raw material or product in the beginning of the process and that raw product may be combustible in itself, but as they further refine the product, many times it becomes even more combustible. They start out with a flammable product, they shape it, grind it, polish it, and create what are called fines or fine dust, what we will call combustible dust in this book. At each point of the process, as the product moves down the line, they have electrical and mechanical equipment or conveyors that can also create heat and friction, and sparks.

Every time they move or manipulate the raw product, they've got the potential for creating combustible dust, and as they're creating combustible dust along the process, they also have the possibility for creating friction and heat. Thus, when you put those two elements together, along with oxygen, this forms the basic fire triangle that many people in my industry are familiar with; where you're going to have oxygen, you're going to have the possibility of ignition, and you're going to have fuel. That combustible dust is the fuel for fire. Heat and friction from the process is the ignition point, that or self-heating, or possibly even mechanical failure and human error from welding, etc., and you typically will have oxygen in most of these processes. You will have these ingredients for fire just through a natural, typical manufacturing process, you have the all the components necessary for fire. Many times, people will say *"well, we haven't had a fire, knock on wood."* Well, they just haven't gotten the recipe right, at the right time!

It may just be a matter of time. Your process isn't static, in other words it is constantly changing and evolving. So a plant, production line, or process that runs for a long period of time without an incident, may over time become more prone to fires as product mixes change, and machinery wears out.

Many times plants run along for years having small fires, easily contained. But one day they have a major incident.

You can do a Hazard Analysis, and analyze your process for combustible dust and ignition hazards, and implement systems and procedures to help prevent these small fires from becoming a major incident. Or we can help you do it.

Chapter 7

The Evidence Does Not Lie

Leading Indicators

A lot of times I'll hear, *"well, we've been running for years this way"* or *"occasionally we'll smell smoke, but we haven't had a fire"*, or *"we hear these hiccups or burps in the system"*, and what they don't realize is those are all precursors to a fire or an explosion. These are all signals or warning signs, leading indicators that something could potentially happen in their process. A lot of times I'll start asking questions, such as *"Have you ever had any fires in that dust collector?"* *"Have you ever had any fires in this machine?"* or *"Have you ever had any sparks or smoke come off that machine?"*, or *"When you go out to your dust collector and you change those bags out, do you ever notice little pinholes or burns in those bags?"*

These are all strong indicators to me that they're having sparks and a potential for fire. Smoke or quantities of sparks are leading indicators that something will eventually happen in this process. They have combustible dust, but they just haven't yet had a spark with the right ignition energy and temperature land in a dust layer, or in a combustible dust cloud that was the right air-fuel ratio to have a fire or an explosion; but that doesn't mean it can't happen, and it probably will happen. This is called luck. And luck is not a proactive strategy for safety.

Chapter 8

Are Fires Just Part Of The Process?

These things happen, but…

We read many times in news stories and reports, where a reporter might interview an owner or a plant manager, employee, first responder, or even the local fire marshal on the site and they'll say something to the effect of *"these fires can be expected, they're just part of the manufacturing process"*.

Well, I disagree. Yes, you do have the ingredients for fire, but that doesn't mean you have to have a fire. Many times, we also see an attitude of complacency where people say *"well, we've always had these little fires, it's just the natural part of our manufacturing"*. Again, maybe it's just luck, or they haven't had significant changes to the product or machinery that would increase the odds, and might actually help create a major fire.

For example, a sugar company ran for 50 years in Georgia, all the while having small problems, but no major incidents. Then they tried to improve part of the process, and enclosed a conveyor to help prevent dust emissions.

There was a fire in the enclosed conveyor that contained sugar and fine sugar dust in suspension - basically a dust cloud. That dust cloud when ignited presumably by an overheated bearing, caused a fire and explosion within the conveyor, which acted like a conduit for the explosion. That explosion moved from one part of the process to the next like falling dominoes, creating a large cascading "secondary explosion", and blew up a major portion of the plant, and ended up killing 14 people, and injuring many others. A security camera across the river caught the whole thing. That catastrophic explosion was seen for miles.

These things can happen. And it is not just a part of the process as many people seem to think. These types of fires and explosions can be prevented, but you have to understand the hazards of combustible dust. It is easy to get complacent, especially when people do not recognize the risk they are working around day in and day out. And when they don't realize how just a small change in the process, or the product mix can have *a magnified effect on risk*. This is why OSHA in their PSM Process Safety Management documents require a change management process be put in place, an MOC Management of Change. That means any time you change your process or product or product mix, you need to re-evaluate the hazards and risk.

Another example is the lumber industry in British Columbia. They were used to processing timber for decades without any major problems. But they had a pine beetle infestation, that basically infested whole forests, and killed the timber. This standing timber dried out before it was harvested. So instead of the "green" timber they were used to with high moisture content, they were now processing pre-dried timber, which had a much lower moisture content, less than 30%, making the timber much more flammable.

When this beetle killed timber was sawn and planned the chips, sawdust, and fines were also much drier and combustible. The operators didn't recognize this. So the sawdust that typically built up around machinery was highly combustible, and dried out further over time, and caused devastating fires and explosions at two of these large mills in B.C. and killed a couple people.

Hierarchy of Safety System Design

We use hazard identification, and a hierarchy of safety systems for risk management called engineering controls, to help prevent these fires and explosions; and, along with your housekeeping program to manage the dust and proper operation of the equipment, proper maintenance of the equipment to prevent that heat and friction, and dust from building up, called administrative controls - these fires can be prevented.

Sometimes the process itself has to be improved or redesigned to help remove or isolate, or redirect the hazard. Substituting inherently safer equipment, or inherently safer process design is sometimes required.

Your Process Is Not Static

Operators tend to think of their processes as being static, unchanging, when in fact manufacturing processes tend to change slowly over time. It could be the raw material they are bringing in may change from that particular manufacturer, or an alternate supplier may have a slightly different product, or the product mix that they're using may change, or the process machinery itself may wear out over time. Sometimes maintenance changes out bearings or router bits, blades, mills or grinders, motors, pulleys and sheaves, they may replace with like type of materials or supplies, but they may not be exactly the same material or size.

The manufacturer may have changed the part, or changed suppliers, so the part or properties of the replacement may not be identical. A buyer or maintenance manager may have swapped for like parts without realizing the implications, without considering change management and hazard analysis, and the implications to safety.

Things Change

These manufacturing processes can and do change over time, especially as parts in the processes wear out and they start creating more friction and heat, and combustible dust emissions migrate from one part of the plant to another. The finer lighter dust often travels and settles in high places such as beams and rafters, creating the perfect storm for a secondary explosion, should this dust become airborne and form a dust cloud.

Even if improvements and changes aren't made, the product mix can change. The equipment and process wears out, also operators are pushed for production, so this attitude of complacency becomes a real problem in some of these older plants, or even as we have seen in new plants as they are pushed to get into production. And I am not implying necessarily that people become complacent with safety, it is just that many times they don't realize the hazards of combustible dust, or changes to the process or product and the possible implications. Or they become overly familiar with safety rules, and maybe a little lax over time. This is just human nature.

The Education Process

The purpose of this book, the articles I write, the presentations I give, and the trade shows we do are all part of the education process. To help educate people to the dangers of the combustible dust they work around every day. So that we can help raise awareness to the dangers of combustible dust, and help protect people and processes.

So, think about these things. Do you currently have any of these problems, or have you ever had any of these issues? Do you think you could have any of these type issues in the future?

And as you think about these things, also think about how we may be able to help you. As an added benefit, we have also provided **Additional Resources and Information** in the back of the book for you.

Chapter 9

Prevention Vs. Protection

Don't We Need Both?

Exactly. You do need both... You need layered fire prevention, and protection systems. By fire prevention, what we mean is we're trying to prevent a fire, by detecting that fire in the beginning stages, before it develops. Then you need to add fire protection systems to put out a fire, as well as explosion protection.

Prevention is like insurance, it's harder to sell the idea of prevention than it is protection. People understand they need a sprinkler system to put out fires, or that they may have a potential for an explosion, so they need explosion protection. It is also a requirement of the NFPA Combustible Dust Standards.

A lot of times operators and managers don't see the need for hazard monitoring and or a fire prevention system until we explain that we're trying to detect that spark or ember, or friction, or heat in the insipient stage, and put it out before it ever starts a fire so they can just keep on running without affecting production. That means less downtime, more uptime, more production, and less risk of damage and injury. And that is the bottom line. More uptime and more production.

Return on Investment for Fire Prevention

There actually is a return on investment to fire prevention. Like I said, if we can prevent these fires, we can prevent the explosions, and we can help you keep running and not hurt people.

See the report by the NFPA on Fires in US Industrial and Manufacturing Facilities that shows from 2006-2010 an estimated 42,800 fires, with 22 deaths, 300 injuries, and $951 million in property damage.

We can help you prevent these fires and losses.

Fires in U.S. Industrial and Manufacturing Facilities

From the National Fire Protection Association
Report: NFPA's "Fires in U.S. Industrial and Manufacturing Facilities"
By Author: Ben Evarts
Issued: April 2012

An overview of industrial and manufacturing property fires, including trend tables, causes, time of day, day of week, month of year, and area of origin.

Abstract

During 2006-2010, an estimated 42,800 fires in or at industrial or manufacturing properties (including utility, defense, agriculture, and mining) were reported to U.S. fire departments per year, with associated annual losses of 22 civilian deaths, 300 civilian injuries, and $951 million in direct property damage. 70% of these fires occurred outside or in unclassified locations, 20% occurred in structures and 9% in vehicles. Two-thirds (66%) of the combined industrial or manufacturing facility structure fires occurred specifically in manufacturing facilities (as opposed to utility, industrial, defense, agriculture, and mining properties).

See:
http://www.nfpa.org/research/reports-and-statistics/fires-by-property-type/industrial-and-manufacturing-facilities/fires-in-us-industrial-and-manufacturing-facilities

---

68

Chapter 10

What Are My Risks?

Your Risk

There are quite a lot of risks associated with fires in many manufacturing plants, and often they are not recognized.

I've had conversations with people over the years at manufacturing plants that did not want to invest in fire prevention, and ended up having a fire or an explosion that killed someone. They got sued. They lost millions of dollars, couldn't operate anymore, and went out of business. A lot of times, they just didn't understand the risks.

As professionals we monitor and read media reports of fires in manufacturing industries on a daily basis. Monthly, we read about explosions in industry and typically annually we'll read about or hear about a catastrophic explosion that kills several people. These incidences, even though they're on the decline in many of our mature industries, some of our newer industries will have more fires and explosions, unnecessarily, and sometimes on a monthly basis.

We have the actual research from NFPA, National Fire Protection Association that also includes some of the research from NFIRS; National Fire Incident Reporting Service, which is a voluntary reporting service for fire fighters. We also have research from FM Global, which is a large insurance company that serves many of these manufacturers. We can actually see how many of these incidences are occurring. We can see what equipment's involved in ignition. We can see the number and dollar amount of losses, and sometimes it's staggering.

These organizations make the information available so we can see how many fires and explosions are occurring in the industry, not rely on just the media reports.

Just recently, we read about several manufacturing plants in China that have exploded and killed people. Some of those manufacturing plants are manufacturing the aluminum parts and components for American consumer products companies. Plants are exploding. People are dying. This threat is real, and not to be trite, but it's a big deal.

The National Fire Protection Association

The National Fire Protection Association provides standards for fire prevention and protection. NFPA 652 provides the fundamental requirements for all industries with combustible dust hazards. NFPA 652 now provides a baseline for all other industries. While NFPA 654 is more general than the other commodity-specific standards, its focus is directed towards chemical and other similar hazardous processing industry. NFPA 654 contains additional requirements that go beyond those in NFPA 652. So these are both good standards to review, as well as your commodity specific standard if applicable.

FM Global also provides data sheets that are good to review, to help protect your process or dust collection systems.

As an added benefit, we have also provided **Additional Resources and Information** in the back of the book for you. If you have any questions about these resources, or any other topics presented in this book, please feel free to contact us at the addresses in the back of the book.

Chapter 11

Who Is At Risk?

Who is at Risk?

Any company that manufactures a product and creates dust in the process, is potentially at risk. Any company that utilizes powders, mixes, dust, fines or chemicals in the manufacturing process is likely at risk - unless they are inert dusts, powders, chemicals or metals.

Although most of our customers are very diligent, many of the industries we work in do not believe they have as much risk as they actually do. OSHA lists many of the industries at risk in their Combustible Dust National Emphasis Program.

The US has had the grain industry standard since the 70s when they had so many grain silo explosions. In 2008, OSHA re-issued their national emphasis program on combustible dust for our industries in the US. The US has been behind the curve. Europe has typically been ahead of the curve with their ATEX standards, but there are many third-world and developing countries, that are "behind the 8 ball" when it comes to safety standards and they're having a lot of fires and explosions, with a lot of injuries and deaths in those countries. China in particular stands out in my mind because of recent events.

Let us help you prevent these types of combustible dust fires and explosions.

Chapter 12

Minimizing Your Risk

How Do We Minimize Our Risk?

Uncertainty in manufacturing is the wild card that can potentially stop production or put you out of business. It is the possibility of an unpredictable and uncontrollable hazard. In this book we are talking about fire or explosion hazards of combustible dust in your process. This risk is the potential consequence of continuous production in spite of this uncertainty.

So, how do we make certain we can minimize this potential combustible dust fire and explosion risk?

How do we maximize production with certainty that we are doing it safely?

Risk Defined

In our industry and in the insurance industry, Risk is defined as the Probability of Occurrence, meaning the likelihood that something could happen, and the Severity of consequence, the magnitude of loss during a possible event.

In other words, you are looking at failure probability multiplied by the potential damage related to the failure. These possible consequences are your risk. The insurance industry will look at property damage. In the fire protection industry we must also take into account life safety.

First is the probability of fire or explosion. So we ask the questions related to probability. The first question is do you normally, or during an upset condition, have the ingredients for a fire? Do you have combustible dust? Do you have possibility of ignition sources? Do you have ignition sources that can cause a combustible dust fire or explosion?

Get Dust Tested

If you do not know if your dust is combustible, you need to get your dust tested. You can get a simple go/no-go test done fairly inexpensively at a regional test lab. If your dust is combustible then you must put in systems and processes for mitigation.

Minimize Risk

How do we minimize our risk? How do we know how bad a fire or an explosion could be, and how much damage it could cause?

We look at risk as probability of occurrence multiplied by the potential severity of consequences of these possible events, including the lives of those working in the immediate area, then we look at how to mitigate.

Reduce Probability and Severity

What we're trying to do in our industry is first reduce that probability of occurrence by utilizing engineered controls for prevention, and then we want to reduce the severity or the magnitude by putting in the various fire protection, as well as explosion protection systems. We're trying to reduce the risk - or the probability of occurrence by using these safety and prevention systems. Hazard monitoring systems such as spark detection systems, are some of the more commonly used tools to prevent the probability of occurrence.

Then we add in layered safety systems, sprinkler systems, deluge systems, fire suppression systems, and explosion protection systems to help reduce severity of consequences of an event. You could actually look at this as a pyramid or hierarchy of controls, with the most effective at the top of the pyramid. But these are the basic, layered safety systems that we add in to help manage that risk.

Engineering Controls
Administrative and Work Practice Controls
Personal Protective Equipment

If we add engineering controls including the hazard monitoring and layered safety systems, then you would also include administrative controls, including housekeeping and change management, to further mitigate risk. And finally add PPE, personal protective equipment to prevent injury.

Change Management

Any time that product coming into the process changes, you should have a change management procedure to verify that product is not going to cause more potential fire hazard in the process.

In our example previously, we saw this in Canada several years ago where two large lumber mills had been running for decades without any major fires or explosions. Then they started processing beetle killed lumber.

They had a wood beetle infestation. This was lumber that the wood beetle had killed whole forests up in Canada. It had dried out as it was standing; this wood had dried out from these beetles having killed the lumber.

Instead of higher moisture content in normal lumber they were bringing in, now they were bringing in this beetle killed lumber that had much lower moisture. This beetle killed timber created a lot drier combustible dust than they were normally used to working with. They didn't have any change management procedure. They didn't realize they needed it until they had two major lumber mills explode with total devastation and killed 2 people.

That's what we mean by change management, and additional administrative controls whether they're housekeeping, keeping the place cleaner, proper operation of the equipment, proper sequences of starting up the equipment and process, and proper sequencing of shutting it down and emergency stop. As well as proper bonding and grounding, and good preventative maintenance practices for equipment.

Those are things that companies can monitor for and institute on their own. As far as the hazard monitoring equipment we use to help manage the risks, we might be using bearing temperature, speed, and run time monitoring; or spark, ember, flame, and smoke detection, as well as possibly CO or combustion gas monitoring in the silo or dust collector, bin, or a particular part of the process. Spark detection systems are one of the easiest and most effective and efficient tools we use for fire prevention.

At the dust collector, we might be monitoring emissions or broken bag detection. From beginning to end of the process and everything in between; spark, ember, flame, anywhere where we can detect something that might create an ignition hazard in the process. That's the type of hazard we're looking for. We use various engineering controls to monitor those hazards, and then we'll add in various types of additional engineering controls to help either divert or suppress that hazard from the process so you can keep running and not lose production.

Who is responsible?

Under OSH Law employers have a responsibility to provide a safe workplace.

https://www.osha.gov/as/opa/worker/employer-responsibility.html

But employees are also responsible for following all of the safety standards and procedures the employer puts in place, however if employees fails to follow the safety rules put in place, the employer will be the one cited for these failures. So safety training is paramount.

In other words, we are all in this together. So everyone must be made aware of the risks involved in your manufacturing process.

Chapter 13

Hazard
Analysis

Site Visit and Hazard Survey

When we make a site visit and survey, we do a quick, very focused hazard analysis and audit of the process itself, where the potential ignition hazards are, what spark producing machinery are part of the process, and where the fire and explosion hazards are.

We will look at all the equipment in the process and ask questions about the history of every machine, conveyor, storage bin, hopper, silo and dust collector. *"Have you ever had any fires in this part of the process? Or at this machine? "Could these mills be creating sparks?"* And *"How about the bearings on these conveyors, ever had any problems with these overheating?"* This gives us an idea of the history and potential for future fires in the process.

Hazard Monitoring

Bearing temperature, run time, and speed for example, is something we would use for hazard monitoring in the grain industry. We monitor bearings in grain conveyors because they can create such high friction and heat, and grain dust is so explosive. We will monitor bearings for temperature as well as the run time and speed because we can actually measure how long these bearings run successfully. When you know the average run time, you can do preventive maintenance to replace these bearings before they wear out and create additional friction and heat and stop the conveyor, or worse start a fire. We might also be monitoring the belt speed, so that you know if a conveyor belt breaks or if it's rubbing and creating friction and heat.

We can monitor drop chutes off conveyor belts for plug-ups as well as for sparks, embers, and flames. We can use water spray at that transfer point to suppress that spark for a just a brief couple of seconds if the process can tolerate water. Or we can divert any hazards out of the system to a fire dump. This is especially critical in enclosed conveyors and bucket elevators containing a dust cloud.

A lot of industries can tolerate a certain amount of water and it's the most economical, so we'll use a highly atomized water spray for suppression if possible. If there's a process that cannot use water, then we will either divert that spark ember or flame out of the process to a safe place using a fire dump or diverter, and then reset the process and keep running, and then you can manage that hazard manually with a fire extinguisher, or by automatic suppression systems on the diverted material, away from production. So there are different engineering controls that we use, it just depends really, on the industry and the process, and what that process will potentially produce, and can tolerate.

Once we've identified the hazard we add some type of automatic response to manage that hazard. Sometimes it might be an inerting gas to suppress a fire or a chemical or foam fire suppressant. May times you can divert, fire dump, or abort the hazard out of the process.

The key is to design a safety system that will detect a hazard in the incipient stage and react automatically and be able to determine the threat level and respond accordingly – without affecting production – in that specific process, or part of the process. The goal is to detect hazards and react to manage the hazard, while keeping production going, unless there is an upset condition.

Quantify the Hazard

Hazard monitoring systems detect and help quantify the hazard. Is it a small hazard that we can just suppress or divert out and keep running? Or is it something that is a serious hazard? Are we seeing temperature rising, or a continuous number of sparks coming through the process or dust collection system that may indicate a machine at the other end is creating sparks, that we can shut down the infeed and help protect the rest of the process from fires and explosions?

The hazard monitoring system can quantity the hazard, and a programmed response can be initiated depending on the threat level. These responses can include alarms for the operators who may wish to activate appropriate interlocks, or it can automatically shut down the infeed of a particular machine, or activate a full blown emergency stop. The hazard monitoring process can be as automated as you need it to be, or left to the operator to activate the appropriate interlocks. Of course we prefer to automate as much of the hazard monitoring and interlocks as possible, to prevent human error, and human risk.

There are various types of engineering controls we can use to deal with the threats depending on the threat level and the process. Our goal is to design a system that will detect and react automatically without effecting production if possible - unless there is an upset condition in the process that needs to be addressed immediately to prevent further hazards or damage, then you would shut down the infeed machinery, and possibly the rest of the process, again depending on the level of threat or risk involved, as well alert operators to the problem.

Chapter 14

Code and Compliance

OSHA

In most of the industries we work in OSHA determines the minimum level of safety required to maintain a safe working environment, facility and process for employees. Or it could be a local AHJ, the authority having jurisdiction. For example, it may be a State OSHA inspector, or a local fire marshal, which is the authority having jurisdiction typically at many facilities, or it could be an insurance inspector, which also might dictate how they reach compliancy.

OSHA is the ultimate authority having jurisdiction in most of these cases and jurisdictions. OSHA has a host of information on combustible dust management on the OSHA.gov website and OSHA reissued their Combustible Dust hazard NEP National Emphasis Program in March 2008, that helps define what the problems are and how companies can start to manage those problems. The big picture is really about hazard awareness. OSHA has been developing their combustible dust standard, but it is not yet finalized, so they rely on the NFPA, National Fire Protection Association codes and standards for enforcement.

Currently, as this book is being written, the insurance industry, local fire marshals, fire protection engineers, and OSHA rely on the NFPA combustible dust standards for determining minimum requirements for protecting manufacturing processes. Additionally, many engineers will also look at FM Global, which is a large insurance company, and publishes their own minimum requirements, and best engineering practices from an insurer's perspective.

Typically, when we look at the National Fire Protection codes, we'll start with NFPA 652, which is the standard on the fundamentals of combustible dust. It helps with understanding of the fundamentals of combustible dust, and it teaches engineers and manufacturers about managing combustible dust hazards.

Some of the key points are hazard identification, combustible dust testing, hazard assessment, hazard management, as well as prescriptive and performance based design.

NFPA

The NFPA codes and standards are commonly called prescriptive design standards. They basically lay out a prescriptive method or model of facility and process safety design, which also allows for performance based design if the engineer can show the authority having jurisdiction that his safety design meets that requirements of the prescriptive code or standard.

NFPA codes prescribe how to protect your processes. Once you have an understanding of the combustible dust and ignition hazards within your process, you'd use a prescriptive or performance based design to start designing in the safety systems we have been discussing, to prevent fires and explosions.

The starting point for safety system design is NFPA 652, which is the standard on the fundamentals of combustible dust, then adding your industry specific document, if there is one. NFPA documents discuss capturing combustible dust and implementing hazardous dust identification, inspections, testing, housekeeping, and the various engineering controls we've been talking about as well as your administrative controls. It also points you in the direction of other industry or commodity specific standards.

If there's a standard for your specific industry, it'll point you in that direction for additional information on how to protect your particular process. For example, NFPA 664 for wood dust, or NFPA 484 for metal dust, and NFPA 61 for bulk agricultural products.

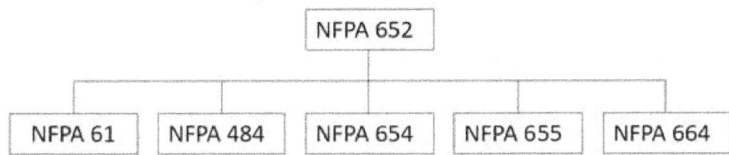

Courtesy, Brian Edwards, PE, Conversion Technology, Inc.

We utilize the NFPA codes and standards, as well as the OSHA documents and FM Global documents to help apply these safety systems for your process.

Really, it depends on how the project is being driven, whether it's being driven by your insurance company or by OSHA, or by your own engineering and safety departments who want to help improve process safety. This is how you determine what the minimum requirements of fire protection, how to meet those requirements, and how to be OSHA compliant.

Below are some typical applicable codes and standards for your review. A more complete list is found in the back of the book under **Additional Resources and Information**. If you have any questions about these resources, or any other topics presented in this book, please feel free to contact us at the addresses in the back of the book.

Applicable NFPA Standards

NFPA 61: Standard for the Prevention of Fires and Dust Explosions in Agricultural and Food Processing Facilities

NFPA 68: Standard on Explosion Protection by Deflagration Venting

NFPA 69: Standard on Explosion Prevention Systems

NFPA 484: Standard for Combustible Metals

NFPA 654: Standard for the Prevention of Fire and Dust Explosions from the Manufacturing, Processing, and Handling of Combustible Particulate Solids

NFPA 664: Standard for the Prevention of Fires and Explosions in Wood Processing and Woodworking Facilities

NFPA 652: Standard on Combustible Dusts

NFPA Standards can be found here:

http://www.nfpa.org/codes-and-standards/document
information-pages

FM Global Resources

7-73 DUST COLLECTORS AND COLLECTION
SYSTEMS

7-76 PREVENTION AND MITIGATION OF
COMBUSTIBLE DUST EXPLOSION AND FIRE

7-78 INDUSTRIAL EXHAUST SYSTEMS

FM Global Loss Prevention Data Sheets can be found at:

https://www.fmglobal.com/research-and-resources/fm-global-data-sheets

Chapter 15

Optimum Safety

Optimum Safety

Optimum safety depends on defining what needs to be protected and at what cost. It often comes down to what the budget can afford, how you can budget for the needed process safety improvements. We work with the customer and use a risk versus reward analysis to determine the amount of safety equipment or systems we can add while still creating additional value, additional levels of safety, staying within their budget.

Some people only look at risk from a dollars and cents perspective, but you also have to take into account life safety, the safety of all the stakeholders, whether they're actual employees operating on the production line, or management, maintenance, housekeeping, or even subcontractors and vendors who come on site. You really have to take into account everyone who's at potential risk in your facility.

If I were in charge of a manufacturing plant and I wanted to have a safe environment for my employees, and be able to sleep at night, I'm going to do everything possible to protect my people. I'm going to add every level of safety that I can, as long as I can see the benefit to safety and maintaining production. I'm going to add many layers of safety and protection for my people, and for production, as production is always the primary motivation in manufacturing. And, if I were in charge, I'm going to go above and beyond to dictate safety standards and procedures, and make sure they're followed by employees and maintenance personnel, and other contractors and vendors coming on site, with severe consequences for not following those safety rules and procedures, just like following any other job requirement or procedure.

It's the attitude of complacency that gets people hurt. This is why you have training, oversight, and continuous improvement of the process, continuous improvement of your safety procedures, continuous improvement of your safety systems to make sure that people do not become complacent. Because in this business, complacency kills.

What is Risk Analysis?

Risk Analysis, is a strategic process of identifying and mitigating risk in the manufacturing process. The typical steps are to define the risk, conduct interviews and walkthroughs. Then assess the hazards, quantify the risk and degrees of risk within the process or parts of the process. Separate acceptable and unacceptable risks. Define preventive measures to reduce the probability of occurrence. Then identify countermeasures to successfully manage these risks when they develop to avert possible consequences that could shut down production.

After analysis, you create path forward action plans, and implement safety countermeasures in order of priority. You may find these terms such as risk analysis and hazard analysis are used interchangeably in your industry.

Hazard Analysis and Risk Assessment

Hazard Analysis is used to identify all possible hazards potentially created by the product or in the process or application. Risk Assessment is the next step after itemizing potential hazards to define and quantify the amount of risk.

FM Global Property Loss Prevention Data Sheet 7-76 for the Prevention and Mitigation of Combustible Dust Explosion and Fire lists the following equipment as being most often involved in loss: dust collectors, impact equipment such as mills, fractioning and sizing equipment, boilers, storage silos, processing equipment, conveyors, ovens, dryers, elevators, etc.

And FM Global Data Sheet 7-76 also lists the most common sources of ignition as: friction, sparks, chemical action, hot work, flame, static, overheating, and hot surfaces.

So for example, if you have any of these types of equipment that can create heat, friction and sparks, such as mills, fractioning equipment, ovens, or dryers, and you have bins and dust collectors on the same process, you would move these to the highest priority for protecting, especially if you have people working around this equipment.

The Risk Matrix

Use a Risk Matrix or spreadsheet to help quantify and prioritize risks within the process. See example below. This gives a project team a quick view of the risks and the priority these risks need to be addressed. You can do this for yourself if you like, without having to hire a high priced consultant or engineering firm. Although, we find that a third set of eyes, and professionals with experience in this field can be of great benefit in helping understand these hazards and how to mitigate them.

First list all the potential dust hazards, and all possible ignition hazards within the process. Then you will need to prioritize and develop an effective strategy.

Creating a Risk Matrix and Plan:

1. Create a list of all likely ignition sources and combustible dust fire risks in the process.
2. Quantify the risk of ignition posed by each, on a scale of 1-5, on the vertical axis of the matrix.
3. Estimate and quantify the severity or impact of each item listed, on a scale of 1-5, on the horizontal axis.
4. Multiply these numbers by each other and list them in order of highest to lowest.
5. Then multiply each number again by the number of people in the area likely to be affected by an incident. This creates a calculation to include life safety.
6. Map out the rankings on the Probability vs. Severity Matrix.
7. Create a Path Forward document to address each risk in order of priority.
8. Develop a response to each risk, based on its priority according to its position in the Risk Matrix, with a deadline for completion.
9. Make someone responsible for managing implementation.
10. Attack the highest priority risks immediately.

		Impact				
		Very Low	Low	Medium	High	Very High
Likelihood	Very High					
	High					
	Medium					
	Low					
	Very Low					

Risk Matrix example

Of course you can get much more in depth with this process, and calculate actual probability, and possible dollars of loss, but it's not necessary for our purposes here. This example is a simple method to calculate the risk of each part of the process, and create an action plan to address each risk, without getting too bogged down in the details.

To create a quick path forward to safety, speed of implementation is key. You can always come back to do a more thorough PHA Process Hazard Analysis or HAZOP Hazard and Operability Study at a later date if needed.

In Part II of this book, we go through a more detailed plan to help you identify and mitigate risk.

Chapter 16

Getting Technical

NFPA Standards are the Place to Start

A "Standard" is a document created by a committee of the NFPA National Fire Protection Association, which contains provisions using the word "Shall" to indicate a requirement. A Standard is used for mandatory reference by another code or standard, or for adoption into law. These NFPA codes and standards represent best practices for industry, and are also cited by OSHA under the General Duty Clause where industry is expected to follow "Recognized and Generally Accepted Good Engineering Practice.

Also know that the IBC International Building Code, the IFC International Fire Code, as well as State Fire Protection Codes also recognize combustible dust as a potentially dangerous hazard, and refer to the NFPA standards. Additionally, the EPA Risk Management Rule, OSHA PSM Regulation, and the OSHA General Duty Clause, while they are not directly related to combustible dust, the implication is that explosions are to be prevented and generally accepted engineering practices are to be followed.

As a side note, all of the NFPA combustible dust standards are continuously evolving and going through revision cycles. They are all now being designed to conform to the outline of NFPA 652, with the eventual goal of having one uniform dust standard which, hopefully, OSHA can then adopt. Although the individual commodity or industry specific standards will likely be annexed or may maintain autonomy.

With combustible dust processing, NFPA 652 is your starting point for compliance. NFPA 652 gives you a framework, an outline for how to address the combustible dust problem. It will walk you through

NFPA 652 Standard on the Fundamentals of Combustible Dust

We find many operators are not aware of the hazards of combustible dust, and many are not aware that the NFPA Standard is actually referenced in fire code, and by OSHA for enforcement. These NFPA Standards as well as other documents referenced in this book including those on the NFPA.org and OSHA.gov websites, and our own website and blog, include many resources for combustible dust education.

The first document you need to become familiar with is NFPA 652 Standard on the Fundamentals of

Combustible Dust, as it provides the general requirements for management and mitigation of combustible dust (also called particulate solids) fire and explosion hazards across all industries and processes, then directs you to NFPA's industry specific standards if you handle wood, metal, sulfur, or grain products. This is a new standard as of 2015, and is the starting place on the hierarchy of standards.

Who is NFPA 652 for?

Anyone who owns or operates a facility where combustible dust could be present, including facilities managers, EHS managers and operations personnel. In addition, anyone who insurers a facility where combustible dust could be present, as well as authorities having jurisdiction (AHJ's), and anyone else responsible for these facilities. Designers and consulting engineers, as well as installer/maintainers and manufacturers of fire and explosion protection and suppression equipment should also reference this standard.

NFPA 652 Brief Overview

NFPA 652 is primarily based on NFPA 654 with additional material and requirements. For instance a DHA Dust Hazard Analysis is now required, and is retroactive.

This NFPA 652 standard will help ensure that fundamental requirements are addressed, such as fuel management (your combustible dust from processing), ignition source control (any fire hazard), and the impact from a fire and/or explosion be limited through inherently safer construction, various types of engineering controls, protection, isolation, as well as your administrative controls.

Your Main Focus in NFPA 652

Your main focus in understanding and implementation of this standard will be hazard awareness, hazard identification, hazard analysis, as well as hazard management involving prevention and mitigation, as we have discussed previously.

NFPA 652 is Retroactive

Some of the requirements in NFPA 652 apply retroactively, including dust hazards analysis (DHA), also called process hazard analysis (PHA) in NFPA 654 and some industries. A dust hazard analysis, or process hazard analysis, is the first step in creating a safety plan for protecting your process from fires and explosions.

Combustible Dust Hazard Analysis

NFPA 652 addresses critically important factors such as DHA combustible dust hazard analysis, dust testing, dust collection, ignition control, as well as housekeeping, and includes mandatory requirements for the management of the fire, flash fire, and explosion hazards of combustible dusts.

Compliance

NFPA 652 will help you better understand the requirements of the NFPA codes and standards, and in turn better understand the OSHA combustible dust requirements for your processing plant, and make compliance with these requirements understandable and manageable.

The NFPA 652 Standard is applicable to all personnel involved with plant operations including management, production, engineering, maintenance, and process safety management and auditing activities, where combustible dust is being generated, processed, or handled.

This standard provides management and supervisors with the insight necessary to identify those hazards with the equipment, operations, processes, and activities that could lead to dust explosions and then help them with the steps that they can implement to ensure compliance and protect personnel, production and profits.

Your NFPA 652 compliance objectives are to determine if a dust fire, flash fire or explosion hazard exists within your facility. It will direct and help you to identify the weaknesses in your existing dust hazard management systems; identify the gaps in your facility's combustible dust fire and explosion prevention and protection requirements; identify the dust samples that need to be tested, the necessary laboratory tests; and conduct a Dust Hazard Analysis (DHA); as well as help you identify practical measures for ensuring compliance.

Commodity Specific Standards

NFPA 652 will also then direct you to your industry specific standard if you manufacture commodities in one of the following industries:

- ➢ **NFPA 61**: Standard for the Prevention of Fires and Dust Explosions in Agricultural and Food Processing Facilities

- ➢ **NFPA 484**: Standard for Combustible Metals

- ➢ **NFPA 654**: Standard for the Prevention of Fire and Dust Explosions from the Manufacturing,

Processing, and Handling of Combustible
Particulate Solids

➤ **NFPA 655**, Standard for the Prevention of
Sulfur Fires and Explosions

➤ **NFPA 664**: Standard for the Prevention of Fires
and Explosions in Wood Processing and
Woodworking Facilities

If you operate in any of these industries you need to comply with the standard for your commodity specific industry. These standards will apply in addition to the NFPA 652 standard. Any conflicts within the standards the commodity specific standard likely takes precedent.

Again these are prescriptive standards, meaning they give you a prescriptive solution to managing hazards in these specific industries. You also have the option for performance based design if it meets the principle of protection outlined in the NFPA prescriptive standards and is approved by the authority having jurisdiction (AHJ).

Part II

The Process

Chapter 17

A Detailed Plan To Manage Risk

A Detailed Plan To Manage Risk

In this, part two of the book I will give you a more detailed plan for how to identify and manage risk, and mitigate combustible dust fires and explosions in your production process. Starting with the basics, and moving to a full blown plan and how to implement it.

Of course these recommendations are general and not specific to your process, all current codes and standards apply. Please refer to the appropriate codes and standards for your process. Before implementation, please get your dust tested and consult a qualified professional about your particular dust, process and situation.

In this section of the book, based on my presentation titled *"Practical Prevention of Combustible Dust Fires and Explosions in Industry"*, I'm going to give you a practical roadmap to show you how to keep from burning down and blowing up your plant.

As the Managing Partner of Industrial Fire Prevention, I've been providing special hazards protection for the combustible dust process industries since 1979. With expertise in protecting process equipment, conveying, fume, and dust collection systems from fires and explosions in many diverse industries.

I started out protecting process and dust collection systems when spark detection and extinguishing systems were first introduced into the United States in the late 1970's. One of the things I'm proudest of is being a technical committee member of NFPA 664 for wood and cellulose material processing.

I believe in continuing education and have studied process safety, fire prevention, spark detection, explosion protection at various companies and organizations including coursework at:

The AAFPA American Forrest and Paper Association
"Understanding and Practical Application of Combustible Dust Hazards in the Wood Products and Paper Industries"

The Fire Protection Research Foundation
"Dust Explosion Hazard Recognition and Control"

Georgia Tech Research Institute
"Process Safety Management"
"Combustible Dust Safety Training
"Preventing and Mitigating Combustible Dust Fires and Explosions"

Additionally, I've written articles related to mitigating combustible dust hazards, for example:

Spark Detection: Plant's First Line of Defence
Understanding the best application of infrared and heat detection sensors is important for effective control systems" Published Q3 2014 in Biomass Magazines' Pellet Mill Magazine

"Spark Detection: First Line of Defence for Preventing Combustible Dust Fires and Explosions" Technical Exclusive Published May 2013 in Powder Bulk Solids Magazine

And Co-Authored this whitepaper, titled:

"Inherent hazards, poor reporting and limited learning in the solid biomass energy sector: A case study of a wheel loader igniting wood dust, leading to fatal explosion at wood pellet manufacturer"
Published by Science Direct in Biomass and Bioenergy Volume 66, July 2014, Pages 450–459

Practical Prevention of Combustible Dust Fires and Explosions in Industry

This section of the book will help raise awareness and understanding of fire and explosion hazards in combustible dust processes, as well as provide a detailed plan on how to mitigate these hazards. We will overview fire and dust explosion fundamentals. As well as discuss general process, fire and explosion prevention principles, and we're going to do a quick overview of the OSHA, NFPA, and FM Global requirements and resources available.

Three key points to take away from this section of the book are:

1) What type of hazard are we dealing with?
2) Why is it a problem?
3) How do we prevent and mitigate this problem?

The First Documented Dust Explosion

The first documented dust explosion that occurred in Turin, Italy in a bakery in 1785. The explosion was caused by ignition of a flour dust by a lamp in the bakery storeroom. It led to the realization that grain dust is a highly explosive substance and that it must be handled carefully.

The Type Hazard We Are Dealing With

Why Is It Such A Problem?

In this picture you see a couple of dust collectors on fire, and this is not just a fire, it's a raging fire. You never want to see this at your facility. These dust collectors go up like Roman Candles. They are hard fires to fight and they put production as well as people's lives in danger.

We monitor media reports for combustible dust related fires and explosions, and just about weekly we see articles on fires or explosions in various industries that use combustible dust in manufacturing or create combustible dust as a by-product of their manufacturing.

These include fires and OSHA citations and fines, plants agreeing to correct their explosion or fire hazards, units that are fighting fires at plants.

Media Reports

- *Industry Plays with Fire and Gets Burned.*

- *Explosion Damages Plant.*

- *Fire at Facility Not Expected to Result in Major Downtime.*

- *Combustible Dust Explosion Prompt Calls for More Oversight.*

These are just some of the type of headlines we see weekly, pulled from recent media reports.

The Ingredients in Your Process That Lead to Fires and Explosions

The first ingredient is fuel, which can be combustible gasses, combustible dust or hybrid mixtures of combustible dust and gasses. Our focus in this book is entirely on combustible dust. Additionally, you may have dust accumulation on surfaces, dust layers. If you have dust emissions from the process, you will have migration of dust from the process to other parts of the facility. You can have dispersion of that dust in air, dust clouds in various types of vessels, and possibly within the plant. You will likely have an oxidizer, oxygen, in sufficient quantity to support ignition. And you typically have ignition sources from friction and heat prevalent throughout the production process.

These are the ingredients needed to create a fire or deflagration. Most of these ingredients are readily available in most manufacturing processes that utilize or create combustible dust in manufacturing.

Fire Protection 101

In Fire Protection, we often talk about the Fire Triangle and sometimes the Fire Tetrahedron. The Fire Triangle requires sufficient Fuel, Oxygen, and Ignition from some kind of energy, heat, friction, or spark to create a fire. You take away any one of those three elements, and you can prevent or stop a fire. If you do have all three of those elements in the right combination at the right time, and they do create a fire, then you have created a chemical chain reaction, which is what we are trying to prevent.

Explosion Protection 101

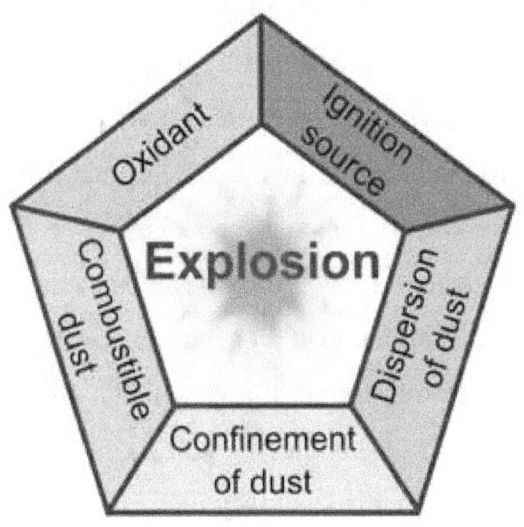

When we talk about explosions, deflagrations, explosion propagation, or explosion protection, we use the Explosion Pentagon as example. Which is the basic Fire Triangle with two additional elements: Dispersion and Confinement. This means you would have sufficient oxygen, ignition potential, and your air/fuel mixture, air and combustible dust in the correct ratio – between the lower and upper explosion limits (i.e. in between too lean or too rich) in suspension (dust cloud), within a confined area or vessel. This is the recipe for a deflagration. And again, if you take away any one of these components, you can prevent or stop a deflagration.

Fire to Explosion Transition

The transition from a fire to an explosion happens when you have oxygen in your process; you're creating or using fuel in your process, in this case combustible dust, and you create heat, friction and sparks in your process that will cause ignition. So, you already have the elements for a fire. And in this example, if you have these ingredients within a vessel or transferred to a vessel with a dust cloud inside, you also have all the ingredients needed for an explosion, or what is called a deflagration.

If you haven't experienced a fire or deflagration, it's simply because you haven't produced those ingredients in the right combination, or the right "recipe" at the right time.

The Flash Fire

Visualize a pile of fine, dry wood dust on the ground or on machinery. It gets ignited either from self-heating, or a spark, ember, flame, or heat of a machine, motor, or bearing. Now that pile of fine, dry wood dust starts smoldering. Well, if you can imagine some kind of a breeze or wind comes across that smoldering pile of fine dust and it bursts into flame. This is how you create a campfire, by taking fine, dry kindling and creating heat, friction, and an ember, then blowing on it to create a flame and fire.

So, imagine that we have that pile of dry smoldering wood dust on the ground, and somebody opens a larger roll up door or larger gust of wind comes across and blows that pile into the air and creates a dust cloud. If it's on fire or that dust cloud has some kind of source of ignition, then it creates a flash fire. If that flash fire happens within a container, a bin, a dust collector, some kind of confined area or vessel, then that flash fire is going to start to over-pressurize that confined space, bin or vessel, because what you have is a flame front that is quickly leaping from dust particle to dust particle at a very rapid rate.

The Deflagration

A flash fire or explosion is basically a fast moving fire, with a flame front, pushing a pressure wave out ahead of it. That pressure wave is thus pushing more fuel out in front of the flame front. As that fireball grows and that pressure wave grows, within milliseconds, it puts outward pressure on the vessel. If that pressure is greater than the vessel can hold, then the vessel will burst. That is what is called a deflagration. As that vessel bursts, the result of that deflagration is what is called an explosion; where you've got the vessel bursting, a flame front, pressure wave and sound wave, you've got shrapnel and a fireball coming out.

That flame front will grow exponentially and with increasing speed as long as there is fuel available. This is why the fireball from the explosion is several times the size of the vessel.

That's how you will have a transition from a smoldering fire, to a fire, to a flash fire, to a deflagration and explosion. This can also lead to secondary catastrophic explosions.

According to Dust Explosions in the Process Industries by Rolf Eckhoff, a dust explosion can be defined as "*any solid material that can burn in air will do so with a violence and speed that increases with the increasing degree of subdivision of the material.*" In other words, the finer and drier the material, the faster it will burn and the more violent it will burn.

Key Point – Fine, Dry Combustible Dust

One of the key points to take away from this chapter is that particle size distribution and moisture content have a serious effect on the propagation of dust fires and explosions, because dust with low moisture content has a lower minimum ignition temperature and ignition energy requirement, and a lower MEC the Minimum Explosible Concentration. We will include this and other key definitions in the back of the book under **Additional Information and Resources.**

As mentioned above, a couple of factors affecting deflagration are the fuel properties: it must be combustible, have a particle size and distribution capable of propagating a flame, and sufficient concentration and moisture content.

You must have dispersion, which is the dust concentration dispersed in air, within explosive limits, the air/fuel ratio. Air and fuel mixtures are defined as having upper and lower and flammability limits, as well as upper and lower explosive limits.

Everything in between is in the combustible range of that air-fuel mixture. You must have ignition and sufficient energy and temperature, and of course, oxygen sufficient to sustain combustion, then confinement, and sufficient over pressurization of the vessel is what causes the deflagration.

Primary and Secondary Explosions

A primary deflagration will often propagate to a secondary process such as dust collectors, storage, and even back to the production area where more dust is dispersed in air and is ignited. The secondary explosion is a result of a domino effect and is typically more violent than the primary explosion and may cause injury, destruction, and death.

Visualize an explosion that happens in a dust collector and realize that flame front is looking for the path of least resistance. If that path of least resistance is in a duct or enclosed conveyor coming back into the plant, then you can have a flame front and pressure wave coming back into the production area.

In this example, that primary explosion caused a shockwave that rocked the building, as you may also likely have fine dust that has accumulated on the rafters, piping, conduit, ductwork, ceilings, walls, machinery, and floors, all that material is now being shaken from its resting place and has now formed a dust cloud in the plant. And as that primary flame front comes back into the plant and ignites the secondary dust cloud within the plant or equipment, it can cause a second or multiple deflagrations throughout the process.

Many times, these are a chain of events, a domino effect. People will report hearing a rumbling like the sound of a train. The secondary explosion is often devastating because it can knock out the walls, knock down the sprinkler systems and fire protection systems, and knock out the emergency lighting. People can get trapped and die in the fire and also from smoke inhalation. These are what are called catastrophic secondary explosions, and are to be avoided at all costs. These are the explosions that kill people. These are the explosions that make the news, and these are the explosions that put companies out of business.

This is what we do, the business we are in, preventing these type fires and explosions. If we can prevent ignition of combustible dust, we can prevent the fires. And if we can prevent the fires, we can prevent the explosions, and the catastrophic secondary explosions.

For more information on how we may be able to help protect your plant from these type fires and explosions, see our contact information at the back of the book.

Chapter 18

"It's Not Rocket Science"

"It's Not Rocket Science"

"The science of explosion control is pretty simple. It's not rocket science," said retired University of Michigan professor of aeronautical engineering, Bill Kauffman. He went on to say ***"If you can see a footprint or write your name on the wall, it's going to explode."***

http://www.insurancejournal.com/news/national/2011/11/15/224119.htm

I like this quote because Bill Kaufman actually is a rocket scientist! Kauffman, a retired Aerospace Engineering professor, expert on combustion and explosions; advised OSHA on its 1987 grain dust rules and acted as an expert witness on a panel that was convened to discuss new combustible dust regulations.

What if Combustible Dust is Just as Flammable and Explosive as Gas?

Think about that. If the combustible dust in your process was gasoline you would not let it pool around running equipment. If this dust was natural gas, you'd shut down the process until you found the gas leak and solved that problem before you restarted the process. It is the same risk with combustible dust. It's just as hazardous as natural gas or gasoline and just as combustible, just as explosive in the right combination. I often show pictures in my presentations of combustible dust and gas explosions side by side, showing the total destruction of each plant.

If you want to prove this to yourself, find pictures on the internet of catastrophic combustible dust explosions, and industrial natural gas explosions, and look at those pictures side by side. You will notice the devastation of each of these types of explosions is the same, in many cases total destruction. The result of a gas explosion and a combustible dust explosion are often the same result.

The takeaway from this chapter is to consider combustible dust just as flammable as gas or vapor.

Why is This Such a Problem?

Carolyn Merritt, the former chairman and CEO of the USCSB, the Chemical Safety Board, was interviewed in June of 2005 by CBS news 60 Minutes program for a program called "Is Enough Being Done To Stop Explosive Dust?"

The interviewer asked her why we are having so many fires in industry. And, her comment was:

> *"Dust has the same potential power of gasoline if a dust explosion occurs."*

The interviewer asked her a follow up question about why it is such a problem, and her response was:

"Operators failed to recognize dust as a critical, catastrophic potential hazard."

http://www.cbsnews.com/news/is-enough-done-to-stop-explosive-dust

At that time, the Chemical Safety Board was investigating a series of catastrophic and fatal dust explosions that had occurred in industry in recent years. In their report of findings the CSB recommended that OSHA needed to develop a combustible dust standard for industry along the same lines as the grain industry standard they had created in the 1980's.

OSHA had a combustible dust program which they re-issued in 2009 and the OSHA NEP Combustible Dust National Emphasis Program states:

"This instruction contains policies and procedures for inspecting workplaces that create or handle combustible dusts. In some circumstances, these dusts may cause a deflagration or other fires or an explosion. These dusts include, but are not limited to, metal dusts such as aluminum and magnesium; wood dusts; coal and other carbon dusts; plastic dusts and additives; bio solids or other organic dusts such as sugar, flower, paper, soap, and dried blood; and certain textile materials."

To summarize, most organic dusts and some metals such as aluminum, magnesium, and titanium, are going to be combustible. Most dusts in manufacturing are going to be combustible. The only ones that aren't combustible are inert ingredients like rock, stone, sand, glass and some types of metals like brass. So, if you've got dust in your process or as a result of your process, you really need to understand the flammable and explosible characteristics of that dust.

The CSB

The Chemical Safety Board found in their study, several common risk factors, including:

- Operators did not recognize combustible dust hazards.
- They had dangerous dust accumulations.
- Their engineering controls were inadequate.
- Their change management controls were inadequate.
- They had inadequate dust collection system design and maintenance.
- They had inadequate fire and explosion prevention and protection systems.

One of the key takeaways from this section of the book is to recognize process dust as a potentially catastrophic hazard.

We are going to outline a series of steps you can take to protect your process. The first of these is to do a process hazard analysis of your dust hazards. Have your dust tested to know if it's explosible and how explosible it is. Know which safety standards apply. Know which electrical classifications apply to your process. Then, you're going to add in ignition control and dust control as well as fire prevention systems and controls. The next layer of protection will be to add fire and explosion protection, and explosion isolation. Then, your administrative controls including preventative maintenance, housekeeping, and change management.

For reference and more information, see the U.S. Chemical Safety and Hazard Investigation Board Investigation Report titled Combustible Dust Hazard Study. This study will show you pictures and give more data as to the hazards involved, Federal Regulations, and Fire Code Enforcement:

http://www.csb.gov/assets/1/19/dust_final_report_website_11-17-06.pdf

Combustible Dust Hazard Recognition

FM Global engineering standards department lists in Loss Prevention Data Sheet 7-76, Table 6, losses by equipment type. They list dust collectors as the number one piece of equipment that are lost due to fire or explosion. They also list impact equipment such as mills, hammer mills, ball mills, and pellet mills; and boilers, storage silos, processing equipment, conveyors, dryers and boilers as most prevalent equipment losses.

Interestingly, dryers account for a small number of losses, but the greatest percentage of financial loss, as most dryer fires and explosions happen during start up or shut down due to improper sequencing of start up or shut down procedures. They also list elevators and spray dryers, bins and other various types of equipment.

FM Global

In FM Global Property and Loss Prevention Data Sheet 7-76 for Combustible Dust Explosions, they list in table 4, losses by cause which are the ignition sources. They list friction, heat, sparks, chemical, hot work, flames, electricity, static electricity, overheating, and hot surfaces.

NFPA

We also have information from the NFPA, National Fire Protection Association data on causes of fires and explosions in industry in their 2012 report titled Fires in U.S. Industrial and Manufacturing Facilities.

http://www.nfpa.org/news-and-research/fire-statistics-and-reports/fire-statistics/fires-by-property-type/industrial-and-manufacturing-facilities/fires-in-us-industrial-and-manufacturing-facilities

See Table 6. Structure Fires in Industrial or Manufacturing Properties, by Equipment Involved in Ignition.

In summary, heat, friction, and sparks are the top causes of ignition the equipment listed above. And dust collectors, dryers, bins, hoppers, conveyors and elevators are the types of equipment losses seen in manufacturing facilities with combustible dust.

If you have these types of equipment and ignition sources, we can help protect your plant from combustible dust fires and explosions. See our contact information at the back of the book.

There is also an offer to send a free copy of this book to someone you know who can use it, on your behalf. See the order page on the last page of this book.

Chapter 19

Process Safety Management

Process Safety Management

How do you keep your process safe or make it safe? First, you do a Risk Analysis. Also called PHA Process Hazard Analysis, and DHA Dust Hazard Analysis. You want to identify the combustible dust hazards in your process, and you want to identify any dust emissions that are coming from your process and may migrate to other parts of the process to create additional hazards.

You also want to identify all ignition hazards in the process. You want to evaluate your process equipment that have the potential for combustible dust fires or explosions. For reference, that would be the equipment we listed in the previous chapter, in FM Global Property and Loss Prevention Data Sheet 7-76, Table 6. Then, you want to implement control measures to mitigate these combustible dust fires and explosions.

Quantify the Risk

How to we quantify risk in a process? Risk is defined as the probably of occurrence multiplied by the potential severity of an incident. You can hire an outside consultant to do this, or you can do this yourself.

Look at the process as a whole, then break it down by process, system, or equipment in the process. Look at individual process equipment, especially the equipment listed in the Factory Mutual tables shown most hazardous or most dangerous, and show the most percentage and cost of loss.

Spark producing equipment will create heat, friction, sparks and are typically mills, dryers, heaters, as well as other process equipment. These are your probable ignition sources.

Your bins, dust collectors, elevators and other equipment with internal dust clouds are your typical equipment involved in explosions.

Rate those pieces of equipment or parts of the process on a scale from one to five as to how frequent a piece of machinery or part of a process can create heat, friction, sparks, embers, fires, and explosions. Rate them from one to five, one being very unlikely and five being that a hazard is very likely to occur. That could be for instance a high-speed hammer mill that's creating a lot of heat, friction, and combustible dust. You want to list those as equipment that has a high probability of occurrence.

If that piece of machinery is connected to a dust storage bin or a dust collector that has a dust could in it, then you want to list it as very high on the severity level as well.

Also look at severity on a scale from one to five. With one being no or low significant risk of injury, and five being likely result in death or serious injury. For example, if somebody was working near a dust collector and it exploded.

The Risk Matrix

You will create what is called a risk matrix based on these findings and scores, see Chapter 18. Then you will create a path forward and list all the high, critical needs equipment or parts of the process that have a high frequency of generating combustible dust and/or ignition sources, or high probability of occurrence; that also have a high probability of severity. And list those with the highest score as the highest priority, needing to be addressed immediately. Speed of implementation of this risk identification process, as well as simultaneous application of safety systems is key to preventing fires and explosions.

Managing the Risk

After identifying and quantifying the risk, you now want to look at managing the risk. You first do the hazard analysis, and quantification of risk, then you want to reduce the probability of occurrence, by adding preventative controls, which we will discuss in detail ahead.

Key: Manage the Dust and Ignition Sources

The Process

To reduce probability and prevent combustible dust fires and explosions, first look at the process design to see if there is anything you can do to prevent fires. By working on preventing dust emissions and ignition, you can make the process inherently safer.

We first look at the possibility of inherently safer design for the facility, the process, and the equipment, then we look at adding fire prevention engineering controls. These controls are actively working to prevent the probability of occurrence of a fire or an explosion. These include ignition control, dust control, and also include your housekeeping and change management in prevention of these dust fires.

Once you have introduced controls and changes to the process to help reduce the probability of occurrence of a fire or explosion, then you want to look at reducing the severity of a fire and explosion should one occur.

These are your protection systems, fire protection systems and explosion protection systems. These are reactive systems. They only react in event of an upset condition. Once you have all this in place, this is what are called layered safety systems. At this point you will have and inherently safer process, with multiple layers of process safety, fire prevention, and protection systems in place.

Inherently Safer Design

Some of the process safety design principles of inherently safer design are to segregate, separate, detach and isolate certain pieces of equipment from each other, or dangerous equipment from one part of a process to the others.

What you are trying to do is prevent a fire and explosion from transferring from one piece of equipment or part of the process to the next. You do that by segregating, interposing a fire and explosion resistant barrier or diverter between the processes; or by separating dangerous equipment or processes from each other, by detaching and isolating, and locating combustible processes in specially constructed areas or separate buildings, or outside the process area.

Diffusing a Bomb

For example, you might have a high speed mill on top of a plenum. On the other end of that plenum, a common design would be to put a dust filter of with a fan on it to pull an air curtain through that bin or hopper to remove the combustible dust from the material.

In this example, that mill is creating a lot of heat, friction, and combustible dust. It's also creating a dust cloud inside of this this bin. At the other end of the bin is also a dust collector with a dust cloud and layered dust inside of it. So, what you've actually created is a bomb!

An inherently safer design would be to remove the dust collector to an outside location, connect it with ductwork so that you're removing the dust to a remote location. Then, you are going to add safety equipment and controls to protect that dust collector.

You will need to add, based on NFPA Standards, spark detection and extinguishing systems typically after a mill, if the process can tolerate water. If not, use some other type of gas suppression or diverter to remove the hazard out of the process.

Also, prior to the dust collector, you will typically need to add a spark detection system on that ductwork and/or on the conveyor. In the dust collector you will add fire suppression - sprinkler, deluge or chemical, so that if a fire does occur in that dust collector it's going to automatically be put it out. According to NFPA dust standards, any duct or conveyor with a fire hazard will require spark detection, and any vessel with a fire hazard will require fire protection.

Then you're going to add isolation so that should a fire and/or explosion occur in that dust collector, it does not blow back into the plant to production, or to the next process downstream, or to storage. You will need to add explosion vents or explosion suppression, and isolation, so that if an explosion does occur, it is controlled. According to NFPA dust standards, any vessel with a deflagration hazard will require explosion protection and isolation.

After reviewing and making any necessary process design changes, and adding in your layered safety system controls above, you will have an inherently safer process.

You've added in your fire prevention system, your spark detection systems, then you've added in your fire protection systems, and your explosion protection systems.

This is a much safer process design. It's much safer for the people working inside around the production line because the main hazard, the dust collector, has been removed outside to a safer location. And you have added in layered safety controls: fire prevention, fire protection, as well as explosion protection and isolation, to help prevent downtime and loss of revenue.

Best Practices

In summary, best practices in process safety methodology are layered safety systems and engineering controls including prevention which are spark, ember, flame, thermal or temperature rate of rise, smoke, combustion gas or CO detection systems, and various types of suppression systems or diverters and fire dumps.

Then add in various types fire protection, sprinkler and deluge systems. This might include water mist, deluge, or dry chemical or $CO2$ and gas inverting systems or fire protection foam. Then add explosion protection which could be explosion venting or suppression as well as explosion isolation systems. Then, interlocking and controls - you may need to add diverters, fire dumps, alarms, deluge, emergency stops, sequential process shut down, with the interlocked response dependent on the threat level detected.

Adding fire prevention reduces the probability of occurrence of fires, utilizing active fire prevention controls. Again, these preventative fire systems are proactive and include spark detection, hot particle detection and extinguishment, suppression and inerting; diverters, flame detection, heat and smoke detection, CO or combustion gas detection, and sometimes we also include--especially in grain industries or other industries - bearing temperature, speed, and run time, as well as belt alignment monitoring because these things create a lot of heat and friction and fires, and explosions as well.

Then, at the other end of the process we can also add in emissions monitoring and broken bag detection in the dust collectors. In NFPA 664, for example, which is the Standard for Prevention of Fires and Explosions in Wood Processing and Wood Working Facilities, talks about ducts with fire hazards or conveyors with fire hazards requiring fire protection. Any duct or conveyor with a fire hazard is going to require some type of fire protection system, typically, a listed spark detection and extinguishing system or deluge and/or deluge system. It could be equipped with a high speed diverter, but this is just an example of one of the combustible dust standards.

Ignorance and complacency kills. Fire and explosion prevention and protection systems save people, process and production. And that is what we do. Let us help you protect your plant.

Chapter 20

Spark
Detection
Saves Lives

Spark Detection Saves Lives

Spark detection and extinguishing systems are a primary source of fire prevention. Spark Detection is one of the go-to tools and engineering controls we use most often in conveying systems, mechanical conveyers, pneumatic conveying, and aspiration, dust and fume collection systems. Designed to detect sparks in the incipient stage prior to ignition, it really is one of our secret weapons in the war on preventing combustible dust fires!

Spark detectors will detect infrared energy within a duct, or transfer chute, and typically use multiple spark detectors across from each other; first to have complete view of the cross sectional area of the duct or transfer chute, and to view through the cascading or transported material from multiple angles, as well as for detection redundancy.

Then further downstream in the process we can add controls to manage the sparks. In some industries, if the process can stand a small amount of water, it will use a quick burst of highly atomized water for spark suppression.

You want a closed end, plug free nozzle that can highly atomize water spray with a fine droplet size; as this works best for suppression, because the smaller the droplet size the more cooling effect it has. You can use less water if you've got a highly pressurized and atomized water spray. Typically, a few gallons of atomized water in tens of thousands of cubic feet of air per minute to the collector is not a problem in many applications, unless it stays on because of a continued threat. But in that case it may be better than a fire. Alternately, if water cannot be introduced into the system, we may use a high speed diverter, and/or chemical or gas suppression.

In a nutshell, these spark detection and extinguishing systems are designed to detect and suppress or divert sparks from the process, without affecting production, unless an upset condition is detected. They can be programmed to provide outputs for alarm and interlocking of the process in an upset condition.

The unique thing about the spark detection systems we use are the ability to keep them simple and just detect and extinguish sparks without affecting production. Or, they can be as sophisticated as we need them to be, counting every spark, monitoring and recording how often they go off on each machine, how much water, when the operator acknowledges the alarm, when the alarm starts and stops, all down the millisecond. So you can start to see trend analysis for production and maintenance. They can also be networked to the plant information system.

Layered Safety Systems

Add Fire Protection

As mentioned previously, and worth repeating, after installing spark detection and/or other possible controls for fire prevention, typically we will add fire protection systems to help reduce the severity of a fire. These systems are reactive and these defensive controls include thermal, temperature rate-of-rise detection, flame detection, deluge systems, sprinkler systems, or water mist and other types of fire suppression systems.

Add Explosion Protection

Then, after fire protection, add explosion protection to reduce the severity and control a deflagration. These defensive mitigation systems include explosion doors, explosion vents and panels, indoor explosion vents which are basically an explosion vent with a flash suppressor on it so that you can control a vented deflagration indoors, as well as explosion detection and suppression and explosion isolation systems. These are to control the deflagration.

Add Explosion Isolation

And lastly, add explosion isolation controls to reduce the probability of transfer of an explosion from one piece of equipment or one part of the process to another. These can be mechanical isolation systems including airlocks, dampers, diverters, gates, and valves, or chemical isolation systems. When used with explosion protection, isolation can help prevent an explosion from blowing back into the plant, causing further damage.

A Fireball from an Explosion is Many Times the Size of the Collector

You must realize, even if we use explosion vents to control the deflagration, that fireball, that flame front coming from that dust collector will be many times the size of the vessel. Based on the explosiveness of your particular dust, create either a barrier or a safe space around your dust collectors, so that if they are in operation and personnel are working around them, they know to stay away a certain distance. They also need to know to wear flame resistant clothing if they are working in the vicinity of a dust collector that's in operation.

Chapter 21

Keys to Process Safety

Keys to Process Safety

You want to control your dust and ignition sources. Then, you want to include prudent engineering controls and interlocks, alarms, machinery shutdown, and sequential process startups and shut downs. Then also add or review your administrative controls such as preventative maintenance, housekeeping, safety procedures, and change management procedures, as well as combustible dust, and fire alarm training for your employees.

Get your Dust Tested

One of the keys to knowing how to prevent and protect against fires and explosions, is to know the explosive characteristics of your dust. When you get your dust tested, there are two types of tests you want to get.

The first one is a Go/No-Go Test. If you have any question, it will tell you whether your dust is combustible or not. You want to get that tested as it comes out of your dust collector, as-is. A general rule of thumb is that if you question whether your dust is combustible or not, it likely is. Get it tested.

If your dust is combustible, you also want to have it more thoroughly tested so you can design your explosion protection systems based on your process equipment and your dust, as explosion protection engineering and design is based on how explosive your dust is.

Some of the things a comprehensive dust test will tell you are your KST, which is your deflagration index. It tells you how explosive your dust is compared to other explosive dusts, and it's an indication of the rate of pressure rise. MIT, your Minimum Ignition Temperature for each a dust layer, and for a dust cloud. The MIE, Minimum Ignition Energy it takes to ignite a dust cloud and dust layer. As well as the MEC, Minimum Explosible Concentration, which is that minimum air/fuel ratio required to create a deflagration.

Recap

So, to recap the last several chapters, some of the keys to process safety are to utilize inherently safer design principles, have your process hazard assessment and dust assessment and testing to understand the potential dust hazards, include training and awareness for operators so they know what kind of dangers they're working around; and so everyone involved knows how to protect the process, the product the production and personnel from fires and explosions.

If you have any questions or concerns with process safety feel free to contact us at the addresses at the end of the book.

Applicable Regulations

Some applicable regulations may include the NFPA National Fire Protection Association, OSHA, OSHA's Combustible Dust National Emphasis Program and their general duty clause, and the EPA's National Air Quality Standards.

OSHA Regulations

On the OSHA.gov website, you can find many, many resources for combustible dust including the Combustible Dust National Emphasis Program, their Safety and Health Information Bulletin, their flyer on Combustible Dust in Industry and How to Prevent and Mitigate the Effects of Fires and Explosions. You can go on the OSHA.gov website and review, read, and download these documents for free.

NFPA Standards

NFPA Standards will be the primary point of reference for all above referenced regulatory agencies, other AHJ Authorities Having Jurisdiction, and insurance underwriters excluding Factory Mutual. FM Global or Factory Mutual has their own set of data sheets that they use. Some states and jurisdictions will also have their own standards.

NFPA, the National Fire Protection Association also provides availability to these standards online, that you can sign up on their website, and read for free. You'll have to pay for those if you want hard copies or if you want electronic copies, but you can go on their website and register and read them for free. Those include NFPA 652, the Standard on the Fundamentals of Combustible Dust, and the other dust documents.

In the hierarchy of the dust standards, NFPA 652 is the top. This is the beginning point for study of combustible dust hazards. NFPA 652 basically outlines what combustible dust hazards are and how to protect your plant. It goes through a lot of the process that we've already discussed.

NFPA 652 will also point you in the direction of industry specific or commodity specific standards such as NFPA 664, the standard for prevention of fires and explosions in wood processing and wood working facilities, or NFPA 484, the standard for combustible metals, or

NFPA 61, the standard for prevention of fires and dust explosions in agricultural and food processing facilities.

If your dust does not fall under any of those standards, you'll also want to look at NFPA 654 in conjunction with 652. NFPA 654 is the standard for prevention of fire and dust explosions from manufacturing, processing, and handling of combustible particulate solids. This is particularly relevant to the chemicals and plastics and other industries that aren't listed in these commodity specific standards.

NFPA 652 Chapters

Some NFPA 652 Chapters you need to be familiar with:

- Chapter 4 General Requirements
- Chapter 5 Hazard Identification
- Chapter 6 Performance-Based Design Option
- Chapter 7 Dust Hazards Analysis (DHA)
- Chapter 8 Hazard Management: Mitigation and Prevention
- Chapter 9 Management Systems

FM Global

FM Global, provides availability to their standards online, called Property Loss Prevention Data Sheets, that you can sign up on their website and read for free at

FMGlobal.com. These include some of the pertinent documents that we referenced in the previous chapters. FM Global Loss Prevention Data Sheet 7-73 on Dust Collectors and Collection Systems, 7-76 Prevention and Mitigation of Combustible Dust Explosions and Fire, and also 7-78 Industrial Exhaust Systems. These are some of the most commonly used in our industry.

We provide a more thorough list of referenced documents for your review in the Additional Information and Resources section of the book.

Again, if you have any concerns with combustible dust process safety, fire or explosion hazards, feel free to contact us at the addresses at the end of the book. We look forward to working with you.

If you know anyone who could use this book, I would like to send them a free copy as a gift on your behalf! See the order form on the last page of this book.

Good luck!

Additional Information And Resources

For More Information

Our Website

For more information on fire and explosion protection for combustible dust processes, you can see our website at: www.industrialfireprevention.com

Our Blog

If you'd like more news and information on combustible dust issues, we compile and consolidate a lot of information about combustible dust, and combustible dust fire and explosion prevention on our blog at: www.industrialfireprevention.blogspot.com

Contact Us

We also have some other books coming out later this year including *The Ultimate Guide to Spark Detection* and *The Ultimate Guide to Dust Hazard Analysis*. You can also contact us at 800-367-0063, or for more information: info@industrialfireprevention.com

Free Gift

If you know anyone who could use this book to help make their process safer, I would be honored to send them this book free of charge, as a gift on your behalf! See the order form on the last page of this book. Any comments are also welcome.

Biography

Jeffrey C. Nichols, Managing Partner, Industrial Fire Prevention, LLC has been providing special hazards protection for combustible dust processes and helping protect production and personnel in the process industries from fires and explosions since 1979.

He is a Technical Committee Member of NFPA 664 the Standard for the Prevention of Fires and Explosions in Wood Processing and Woodworking Facilities.

Mr. Nichols has undertaken coursework in Preventing and Mitigating Combustible Dust Fires and Explosions, Combustible Dust Safety Training, as well as Process Safety Management of Highly Hazardous and Explosive chemicals at Georgia Tech Research Institute. As well as Process Safety and Industrial Explosion Protection from StuvEx, Explosion Protection Fundamentals at Fike Corporation, Dust Explosion Hazard Recognition and Control from The Fire Protection Research Foundation, and Understanding and Practical Prevention of Combustible Dust Hazards in Wood products and Paper Industries from The American Forest & Paper Association. He has also written several articles on spark detection for various publications.

With expertise protecting various types of process equipment, conveying, fume and dust collection systems from fires and explosions in many diverse industries, he started protecting process and dust collection systems when spark detection & extinguishing systems were first introduced into the United States in the 1970's. Having helped install, test, and refine the first spark detection and related systems for North America, used in the protection of dust collection and related process equipment, over the years he has also accrued expertise in applying a hierarchy of other hazard monitoring, fire and explosion protection systems, as well as combustible dust consulting and training.

Other Articles by the Author

Publications:

Spark Detection - First Line of Defense for Preventing Combustible Dust Fires and Explosions
Article published April 9, 2013 Powder and Bulk Solids magazine.

Spark Detection: Plants First Line of Defense – Understanding the best application of infrared and heat detection sensors is important for effective control systems. Article published Q3 Pellet Mill Magazine.

Wheel loader ignites wood dust at pellet mill manufacturer leading to fatal explosion - the incident highlights inherent hazards, poor reporting and limited learning in the sustainable biomass energy sector.
Whitepaper published in Biomass and Bioenergy Volume 66, July 2014, Pages 450–459.

Recent Featured Guest Speaker at the following conferences:

➢ Wood Bioenergy Conference & Expo 2016

➢ PELICE Panel & Engineered Lumber International Conference & Expo 2016

➢ National Safety Council Congress and Expo 2015

➢ International Biomass Conference and Expo 2014

➢ Georgia Safety Health, Safety and Environmental Conference 2014

➢ OH&S Occupational Health and Safety Conference 2013

➢ Tennessee Health and Safety Conference 2013

➢ Bioenergy Conference & Expo 2012

➢ TAPPI BioPro Expo 2011

➢ NFPA Americas' Fire and Security Expo 2009

Key Definitions

Deflagration Key Words
- **Kst** - Deflagration Index
- **MIE** - Minimum Ignition Energy
- **MEC** - Minimum Explosible Concentration
- **MIT** - Minimum Ignition Temp – dust cloud, Minimum Ignition Temp - dust layer
- **LOC** - Limiting Oxygen Concentration
- **LFL** - Lower Flammability limit
- **Pmax** - Maximum Explosion Pressure
- **(dp/dt)max** - Max Rate of Pressure Rise

Ignition Sensitivity
- **MEC** Minimum Explosible Concentration
- **LEL/LFL** Lower Explosible, and Lower Flammability Levels
- **MIT** Minimum Ignition Temperature
- **MIE** Minimum Ignition Energy

Explosive material/substance

Those capable of causing an explosion influenced by confinement.

Hybrid Mixture

A mixture of a flammable gas with either a combustible dust or a combustible mist.

Minimum Explosive Concentration (MEC)

The minimum concentration of combustible dust suspended in air, measured in mass per unit volume that will support a deflagration.

Catalogue of Additional Resources

OSHA.gov Resources

www.OSHA.gov

> - Directives CPL 03-00-008 - Combustible Dust National Emphasis Program (Reissued)

http://www.osha.gov/pls/oshaweb/owadisp.show_document?p_table=DIRECTIVES&p_id=3830

Safety and Health Information Bulletin SHIB 07-31-2005

> - Combustible Dust in Industry: Preventing and Mitigating the Effects of Fire and Explosions

http://www.osha.gov/dts/shib/shib073105.html

NFPA National Fire Protection Association

www.NFPA.org

http://www.nfpa.org/aboutthecodes/list_of_codes_and_standards.asp

> - **NFPA 61**: Standard for the Prevention of Fires and. Dust Explosions in Agricultural and Food Processing Facilities
> - **NFPA 68**: Standard on Explosion Protection by Deflagration Venting
> - **NFPA 69**: Standard on Explosion Prevention Systems
> - **NFPA 70**: National Electrical Code
> - **NFPA 91**: Standard for Exhaust Systems
> - **NFPA 484**: Standard for. Combustible Metals

- ➢ **NFPA 652**: standard on The Fundamentals of Combustible Dust
- ➢ **NFPA 654**: Standard for the Prevention of Fire and Dust Explosions from the Manufacturing, Processing, and Handling of Combustible Particulate Solids
- ➢ **NFPA 655**, Standard for the Prevention of Sulfur Fires and Explosions
- ➢ **NFPA 664**: Standard for the Prevention of Fires and Explosions in Wood Processing and Woodworking Facilities

FM Global Resources

FM Global Loss Prevention Data Sheets
http://www.fmglobal.com/FMGlobalRegistration/Downloads.aspx

- ➢ FM Global Data Sheets 7-0 CAUSES AND EFFECTS OF FIRES AND EXPLOSIONS
- ➢ FM Global Data Sheets 7-10 WOOD PROCESSING AND WOODWORKING FACILITIES
- ➢ FM Global Data Sheets 7-17 EXPLOSION PROTECTION SYSTEMS
- ➢ FM Global Data Sheets 7-73 DUST COLLECTORS AND COLLECTION SYSTEMS
- ➢ FM Global Data Sheets 7-76 PREVENTION AND MITIGATION OF COMBUSTIBLE DUST EXPLOSION AND FIRES
- ➢ FM Global Data Sheets 7-78 INDUSTRIAL EXHAUST SYSTEMS

And Now A Word From Our Sponsor

I would appreciate your feedback. If you enjoyed this book and found it useful, would you mind sending me a few words on what you thought about the book?

_____

_____

_____

_____

Also, if you know anyone who could use this book to help make their process safer, I would like to give this book to them as a free gift <u>on your behalf</u>. So right now, would you please fax me (**770-266-7223**) the name and address of someone that would benefit from this book, as a gift from you.

Please send a copy of your book as a gift from me to:

Name:_____

Company:_____

Address:_____

City:_____

State:_____Zip:_____

PLEASE FAX THIS BACK TO 770-266-7223 – Thanks!
I look forward to working with you!

Industrial Fire Prevention, LLC – Toll Free 1-800-367-0063

www.ingramcontent.com/pod-product-compliance
Lightning Source LLC
Chambersburg PA
CBHW070319190526
45169CB00005B/1670